The Other Side of Steam

J.K. Gourley

The Other Side of Steam

Copyright © J.K. Gourley 2016

All rights reserved.

No part of this book may be reprinted or reproduced or utilised in any form or by any electronic, mechanical, or other means, now known or hereafter invented, including photocopying and recording, or in any form of storage or retrieval system, without prior permission in writing from the publisher.

If you would like to share this book with another person, please purchase an additional copy for each recipient.

ISBN: 978-1523608645

Formatting by Rebecca Emin

Acknowledgements

I am grateful to James Crockett and Roger Thistlethwaite, friends and work colleagues for reading the manuscript and offering useful suggestions. My journalist son Perry for advice, correction and launching the project and finally, but not least, my social worker wife Elizabeth for encouragement.

Introduction

THE intention of this book is to give the reader a glimpse into the engineering insurance industry, particularly the engineering inspection work carried out by many of its employees.

Most of this inspection work is carried out at industrial premises in direct response to the requirements of legislation with certain items of plant deemed to be potentially particularly dangerous, such as steam boilers and cranes. However, a far wider range of pressure and lifting plant also comes under inspection requirements.

The general public will seldom encounter this type of plant, although they will have probably ridden on shop lifts, admired restored steam locomotives, seen building site cranes, or benefited from a hospital steriliser. All these different types of plant will have been subject to periodic inspection by an insurance company employee, generally known as an engineer-surveyor.

The format of the book is part-autobiographical as the human side of the job adds interest to those not normally interested in technical subjects. I hope the emphasis on the interaction between surveyors and clients together with a little humour will dissipate the notion that engineers live within machines in a private, emotionless world!

My use of the word client requires explanation, strictly speaking the client or customer is the one who pays for our services and we act in his best interests which always involves the delivery of a first class service and the best technical advice regarding his plant. Most of my time however was spent with the client's employees, who for the sake of brevity, I also describe as clients. The relationship here is slightly different and on a few occasions within the book this difference will be noticeable!

I spent over thirty five fulfilling years in the job and this is my qualification for writing on the subject. My involvement was with steam boilers, pressure vessels and heavy mechanical plant; there are other surveyors who inspect cranes, lifts and electrical plant who will also have had an interesting career.

Interspersed with my personal experiences are a few chapters devoted to solely technical matters which I hope will be of interest to readers. Finally I must acknowledge that the legislation and the technical side of the job have moved on but I am sure that the enthusiasm and dedication of current surveyors is unchanged.

So why not join me on a tour of inspection? No need to don a boiler suit and work up a sweat and claustrophobia can be forgotten. Be an armchair surveyor - no chance of digging out a clinker from your belly button!

Contents

Introduction

1. Nature and Nurture	9
2. Induction	17
3. History of the Engineering Insurance Industry	26
4. Spreading Oil upon the Waters	30
5. Embarrassment	32
6. Reactors and reactions	34
7. No Fire without Draught	37
8. Claims and Confidentiality	39
9. Repairing the House of Cards	41
10. Industrial Central Heating Boilers	45
11. Plates and Ladders	50
12. Problem Piping	51
13. Safe Landing	55
14. Take me to the Cleaners	57
15. Hydraulic Tests	59
16. Economisers and Ethics	65
17. Maintaining the Circle	69
18. Locomotive Type Boilers	73
19. Boiler Makers - Industrial Craftsmen Par Excellence	78
20. Off the peg – On the Blood	81

21. Vacuum on the Rampage	84
22. Cracks and Fractures	87
23. Rat up a Drain Pipe	91
24. Fittings and Supplies	95
25. The Flying Saucer	99
26. Candle to Calculus	101
27. Temple of Meditation	103
28. Payment in Kind	106
29. Convents and Semantics	108
30. Hands on – Hands off	110
31. Dyeing, defaulting and deception	114
32. Blow-down and bafflement	116
33. Be careful where you sit for lunch	118
34. Brand new – only fit for scrap	120
35. Jacketed Pans and exotic hams	122
36. The Thin Yellow Line	125
37. Trainees, Traumas and Tantrums	127
38. Two Sea Lawyers meet up for a chat	130
39. Finished with Engines	133

1. Nature and nurture

FOR me, engineering is definitely in the blood. My father, a factory inspector, was also an engineer and four of my uncles had a technical background, so from the age of five my genes were unstoppable.

Meccano was the catalyst, out of the hands of the 'Babe and suckling' came forth various types of crane, excavators, gear boxes, engines and many different mechanisms.

Thereafter the passion widened to steam. At eight years old, I had a small boiler supplying steam to an oscillating steam engine. Later, by means of an elastic band, I coupled it to a tiny dynamo which powered a small torch bulb, in my imagination I was superintendent of my own power station!

The very old boiler was of copper, about 5" diameter with flat unstayed ends, secured by swaging over the shell and soldering. The safety valve was tiny compared with the heat output of the gas burner on the stove which must have been equivalent to at least 2 kilowatts - I clearly had much to learn about the safety of pressure vessels!

My next big adventure with steam was at grammar school, where I successfully persuaded the woodwork master (or teacher in today's parlance) to let me opt out of the exciting prospect of constructing a pencil box and build a model boiler instead.

This was to be again 5" diameter, vertical, and about 12" high with a centre flue and coal fired. Made from copper, the seams were of rivets and solder. It turned out to be quite presentable and worked well and at the term end my boiler got a place in the display cabinet next to the best pencil box in year!

The school I attended put great emphasis on the classical subjects and less on the sciences, I barely survived the former but did quite well in the latter.

Looking back over the years I now realise that my placement there was a strategy on the part of my father to rid me of the engineering bug.

My father saw engineering as a lot of effort for poor pay, in retrospect he was correct, few people join the profession with the object of making their fortune. The pay may not be wonderful and certainly is not an incentive, but for most of us the job satisfaction far and away exceeds any monetary rewards.

I left school and despite my parents' persuasive efforts to steer me clear of engineering, I started a five year apprenticeship in mechanical engineering. I lived in Wallasey at the time and this involved getting up at 6.00 a.m. and catching the bus to Seacombe ferry station in order to board the ferry boat for Liverpool, then a short walk to work.

I didn't realise it at the time but this was a huge social change for me. Lunch was taken sitting on a lathe bed, suitably insulated with wooden slats, protection from the dreaded piles! Prior to lunch a fitter's mate, otherwise known as a labourer, would take orders for the banquet. The extensive menu was based on hard cheese of unknown provenance clamped between great slabs of bread. Plain cheese was an option, but not recommended. Cheese and pickle was very popular as it overpowered the disgusting tasting cheese.

There was also an option of cheese with onion or celery - fine during the early part of the week but by Wednesday the vegetables had turned dark brown. The feast was served from a cardboard box by the same labourer. Each diner received his platter carefully wrapped in a copy of the previous day's The Sun newspaper.

It was certainly a different experience from lunch at school where lunch was taken in the 'Refectory' Dining rooms and canteens were presumably for plebs! Standing adjacent to the bench seats we awaited the arrival of 'Sir' duly decked out in gown and mortarboard. A lengthy grace was then articulated in

Latin by the master, ancient language used presumably to demonstrate his scholarship rather than his allegiance to Christianity.

Later in my apprenticeship I learnt what real 'Grace' was all about, not from a scholar but from a humble engineering labourer. I soon discovered that my 'Posh' London accent was somewhat disadvantageous in the land of the 'Scouser' but despite my best efforts I never mastered the new language! Notwithstanding the social challenge, I thoroughly enjoyed every day of the apprenticeship together with three evenings a week at night school learning the theoretical side of engineering.

Halfway through serving my time I spent a period preparing various pressure vessels including steam boilers for inspection. It was during this period that I met the engineer- surveyor who actually carried out the statutory inspections of these vessels.

His job instantly appealed to me, the technical challenges, the rapport (or sometimes otherwise!) with clients, writing technical reports and the freedom to work from home and plan each day's work.

Writing this many years later I can still say that every single day spent at the job was magic, not work, but the greatest pleasure. I cannot imagine that any other job would have suited me so well and completely - how many can boast that in today's work places?

However, before I could even apply for the job, the apprenticeship and night school courses had to be completed and thereafter a period of sea-going for further qualifications. When I eventually got the job as an engineer-surveyor I found that the firm only employed ex-marine engineers in that capacity, some two hundred and fifty in total.

To me the experience of sea-going was the equivalent of attendance at a Swiss finishing school, it really put a polish to my experience and competence. Marine engineering is all about heavy plant, boilers, various types of engines, compressors, generators, pumps, heat exchangers and so forth.

The job entailed watch keeping, maintenance and repairs, but the best part was working with fellow engineer officers, who to a man were enthusiastic, highly motivated and competent.

The machinery on a ship is rarely renewed or upgraded, it remains unaltered for the life of the ship. At times the safety of the ship and its crew depend on it, hence there is a certain affinity between men and machinery seldom found in shore jobs.

I recall an amusing incident which occurred on one ship, I had just been promoted to third engineer and in charge of the twelve to four watch. On this particular morning the main engine, boilers, generators and auxiliary machinery all appeared to be running normally. I was completing the engine room log at about 3.30am, feeling very pleased with myself.

The log book was on a high wooden desk and whilst inserting the bits of data I glanced down and saw a pair of carpet slippers. Moving my gaze upwards, I saw pyjama bottoms then a uniform jacket surmounted by a turkey red face, neck veins fully inflated; it was none other than the chief engineer.

He thundered at me: "What's wrong?"

Looking at him (and bearing in mind it was 3.30 in the morning when chief engineers are seldom seen), my first thought was that he was slightly inebriated and had perhaps been to a fancy dress party with the passengers.

I was contemplating a humorous reply, concentrating on his lack of trousers but the red face and neck veins suggested an alternative approach was perhaps wiser.

Perhaps, unnoticed by me, there was something seriously wrong. With rising alarm I scanned the battery of gauges and saw nothing abnormal. He then strode over to the main engine condenser vacuum gauge and tapped it with a knuckle.

The pointer immediately dropped and indicated a

much reduced vacuum. With considerable haste I speeded up the main circulating pump, the vacuum quickly returned to normal and the main engine slowly increased speed.

The chief then departed from the engine room with a wonderful rebuke to the effect that he was a better engineer asleep than I would ever be awake. I had hoped to repeat this admonishment on some poor soul but the opportunity never came!

The explanation for his appearance that night was that he had been awakened by the slowing engine - a potentially serious problem - so throwing on his uniform jacket had rushed down to see what the problem was. This was a lesson I have never forgotten: don't just rely on gauges and instruments but listen, feel, smell and be vigilant.

The technical explanation for the incident was that the sea water temperature had risen - unnoticed by me - and caused a decrease in condenser vacuum to which steam turbines are sensitive.

This had caused a reduction in power output, hence the slowing down. The ship was either going to the States or returning to the UK, I cannot remember which, but either way it was crossing the Gulf Stream, which caused a rapid fluctuation in sea water temperature hence a reduction in condenser vacuum.

Incidentally, the ability to sleep and at the same time be alert to machinery problems is not unique, the engine attendant at a local textile mill was invariably asleep in his office next to the engine whenever I called.

He could be instantly awakened by me rhythmically lightly tapping some remote part of the engine. With barely time to hide the hammer, he would be at my side enquiring if I had heard the strange tapping noise!

The next ship that I sailed on was very nearly blown apart after a simple mistake was made by a boiler fireman acting with good intentions but without authorisation.

The main boilers were oil fired, supplied with boiler

fuel oil at about 100 psi and at nearly 200° Fahrenheit. A steam driven pump delivered the hot fuel to a duplex filter unit and either filter could be isolated for cleaning purposes.

When cleaning was required the firemen were authorised to isolate the dirty filter by the valve arrangement, drain the unit by a small valve at the base, then unbolt the top flange and lift out the filter element by the overhead chain block. After cleaning, the unit would be reassembled. While rather different from the oil filter change on your car, the procedure had been carried out for years with no mishaps.

On this occasion the ship was tied up in Montreal and in the early hours of the morning the duty fireman had cleaned and reassembled one of the filters. He had noticed a slight oil leakage from the gland of the associated shut valve and had decided to tighten up the gland. Despite doing so, the oil continued to leak.

Because the valve was in the shut position he incorrectly assumed that the gland was not pressurised, although the fact that leakage was taking place should have made him think twice about the matter (hindsight is a wonderful thing!)

He decided to correct the fault by repacking the gland, a procedure for which he was not authorised. The first step was to remove both gland nuts, then remove the old packing and then fit new packing followed by replacement of the gland and two securing nuts.

He didn't get very far with his efforts, no sooner had the nuts been removed when the gland and all the packings were expelled with explosive force, followed by jets of scalding hot oil.

The drop in oil pressure caused the steam driven pump to race, pumping great quantities of hot oil into the boiler room, most of which broke down into a heavy oily vapour, within seconds the hot vapour had reached deck level and was spewing out through various gratings. The burnt fireman attempted to reach the pump with the intention of stopping it but was driven back by the hot vapour. Fearing an explosion he next tried to reach the boiler fronts with the intention of extinguishing the burners but again was driven back.

Finally, racing up the engine room steps he hammered on the cabin doors of the seven engineers, screaming and shouting for assistance. Within seconds we were racing down the steps, barely awake and in a semi-naked state, to find that the vapour was seeping into the engine room and it was impossible to get to the boiler fronts. Like many ships the engine and boiler room were separated by watertight doors which were promptly shut.

We next raced back up the steps to the main deck level from which - like most ships - lever operated trip wires extended down to important valves in the engine and boiler rooms. With incredible speed we pulled every lever we could find in the hope that at least one would shut down the fuel supply and all the boiler room ventilators were turned to face the wind with a view to dispersing the oily vapour.

No sooner than this was done the ship's lights started to dim, which meant that the fuel had been cut off and the boiler burners extinguished, lessening the chance of explosion.

The ship's electrical services were supplied by steam driven generators which were now slowing through lack of steam and threatened to cause a total blackout - this is seagoing, plenty of excitement!

Fearing the worst, I and another engineer re-entered the engine room with a view to starting up the emergency generator, its engine about thirty years old and last test run some three months previous.

The engine had a compressed air starting system so no danger of a flat battery and it fired up instantly and though of limited output it did as required and restored the engine room lighting. After several hours the vapour dispersed, the boilers were relit and slowly all systems returned to normal. Every surface in the boiler room was coated with treacle-like fuel oil and took weeks of overtime working by the firemen to clean. The fireman who had been responsible for the incident was very lucky to have escaped with minor burns and not to have been fatally burnt.

Not all excitement at sea was of a technical nature. On one occasion when the ship was about to sail from a Canadian port. As I was off watch I idly observed preparations for departure.

The mooring parties were at their stations fore and aft preparing to cast off the hawsers and the gangway was about to be lifted when a taxi speeded to the base of the gangway and its three passengers raced up onto the ship, one carrying a large case, the second assisting (or rather man-handling) the third up to the deck.

No sooner had the first two returned to the taxi than the gangway was lifted, the hawsers freed and we were off down river, I was to learn later that our new passenger was a mental patient on his way to another hospital.

The patient, no doubt distressed at his predicament, made his way up to the bridge and armed with a lump of wood threatened the captain and demanded that the ship be returned to port. Before long he was raining blows on the captain who had no option but to flee the bridge.

There followed a chase round the ship, hilarious from one angle but pretty serious from another, and eventually our beleaguered captain found safe refuge in the seaman's mess room. His rescue was led by the bosun and a party of seamen. Eventually the patient was locked into the ship's sick room and remained there for the four-day voyage, fed and watered through the porthole.

It appeared that the authorities had asked the ship's agents if we would take the patient to the next port, conveniently forgetting to mention that the unfortunate man had a mental illness.

***

2. Induction

MY intention was always to remain at sea until I had obtained the necessary qualifications for my move into the engineering insurance industry. When the day finally arrived I scanned the papers for my next and final job.

At the time there were two major engineering insurance companies advertising for staff. Both had a history of industrial plant inspection extending over a hundred years; one had a job in Bolton, the other in Scunthorpe. Both offered the same pay and conditions of service, but somehow the latter did not appeal to me and just sounded a bit down at heel (apologies to the good folk there!).

The three month induction or training period was carried at the firm's Manchester Head Office together with several weeks of field trips to the surrounding towns. It was a pretty thorough and demanding induction. I had my own office, but it was a cubbyhole in the basement with just a chair, table and a light - prison might have been better, after a few days a second chair appeared and I was joined by a second trainee.

Our mentor was none other than the assistant chief engineer of the boiler department, who dealt with all pressure vessel claims. He was on the point of retirement but had a vast knowledge of his subject. He was extremely busy but managed to spend about two hours each day with us.

The rest of our days were spent reading through great volumes of accident reports going back about fifty years and more, yellowing with age. After a few weeks this activity became rather tedious and most mornings we escaped into Manchester armed with a street map and methodically walked the city which was completely new territory for both of us, returning to our 'cell' in time for lunch.

Our mentor, being a Scotsman of the dour kind, was most economical in humour and extremely precise in words and movement. When explaining a report to us, the form had to be first manoeuvred exactly to the centre of his desk and smoothed flat before any exposition (he must have had hidden dividers and feeler gauges!).

Every heading, sentence and word was carefully explained for each statutory form question, together with the correct response. At about quarter to five, whilst the rest of the office were starting to leave, we were kept working till the minute hand reached the precise vertical position, just in time for me to miss the fast train for home!

Under his excellent tutelage we received a good grounding in the statutory and technical aspects of the job, although a year or so into the job I remember being caught out by his exacting and precise method of working.

Many clients required multiple copies of reports, each requiring my signature and I decided to speed this process up by having a rubber stamp made. This proved to be a great time-saver. However, unbeknown to me, my exacting mentor had been taking a keen interest in my report signing and had noted that in every case a French grave accent appeared above the letter 'e'.

In due course I was called into the office to explain how I came to omit my French background in the job application. I blamed the rubber stamp maker for carelessly leaving a sliver of rubber over the letter 'e' but my explanation did not wash too well.

For the next ten minutes I was lectured on the difference between a signature and a stamping and the legal implications, all very sobering.

I was told to destroy the stamp and revert to signatures and from that day on dutifully signed all the copies for clients. However, I continued to use the stamp for internal paperwork (after first removing the grave accent) for the next twenty years – what's wrong with being a rebel occasionally!

Anyway, back to the training and after the period spent in the office came the time to go out with various surveyors. The intention was to observe how the job was done, from meeting the clients, carrying out the inspection and finally handling the paperwork.

It was intended that most of the inspections would be carried out on steam boilers, which although only representing a small amount of plant inspected, do set the scene for inspections of other types of plant.

As an ex-marine engineer this was familiar ground technically, but the relationship with clients, the legal implications and the form filling was new.

I spent several weeks with a surveyor from Birkenhead whose district centred around the town, but owing to difficulties in finding suitable plant for inspection during the winter months we were obliged to travel considerable distances by bus and train (cars came later).

The first job was in St. Helens, a journey of nearly two hours, where we were to look at what turned out to be a small watertube boiler in a very run down factory.

On arrival we found the boilerhouse enveloped in a dense fog of dust as the cleaners were still hard at work. None of the fittings had been taken apart and the place was a shambles. The works' engineer tried to explain that all would be ready in about one hour and would we care to wait in the canteen. My forthright companion hit the proverbial roof. No, we would not wait even a minute and inspection could be forgotten so far as he was concerned.

With that we left the premises and in two hours' time were back in Birkenhead. I had seen nothing of technical interest but at least I had learnt the importance of taking a hard line with the disorganised and indolent. I had acquired a superb role model and as part of the training I had to produce a letter explaining our actions.

The next day we were at a paint factory in Liverpool,

this time it was a small vertical boiler, reasonably clean but none of the joints had been removed from the hand hole covers and no fittings were apart despite a request on the previous report.

The client offered to complete the preparation but his manner suggested a complete lack of urgency which my new hero had obviously detected as he quickly said goodbye and headed for the Birkenhead train.

This was the pattern for the next fortnight, I didn't actually touch a boiler yet alone make any kind of examination. On the final day we attended at the shipyard in my mentor's own district to inspect a boiler on a steam crane. The preparation was perfect, every part was thoroughly clean, every fitting apart and laid out in an orderly fashion.

My companion carried out his inspection and made notes on the defects, I then examined the boiler and did likewise and we compared notes afterwards. Let's just say I had plenty to learn.

We then reinspected the boiler together and went through the reasons for the defects appearing at particular places, followed by a discussion as to their relevance to safety,

It was a really good morning's work as far as I was concerned. I had learnt the standard of preparation required of clients to enable the inspection to be carried out quickly and thoroughly and also had picked up valuable technical knowledge.

My next two trainers had the same degree of technical competence but were as different as chalk from cheese when it came to the personal side of the job. Most of every day was wasted waiting for clients to complete the preparation of plant or in some cases making return visits to complete inspections.

After working all day on inspections, the evenings were spent report writing, much to the annoyance of their wives, to my mind a completely ridiculous situation. Considering their sea going background I was astonished at the lack of leadership and failure to tackle time wasting.

It was all too obvious that the clients had no respect for them and so they made their weary way through life, working all the hours that were sent and finding no enjoyment in the job at all. I was quite determined that on completion of my training I would never work in that way - a mutually agreed date and time would be set, any delay over five minutes or incomplete preparation of plant heralded my immediate departure, not to be seen again for at least a week – not good for popularity, but working wonders in reducing my working hours!

With accompanied inspections completed I was let loose in the world of industry, supervised from the office by my mentor. The plan of action being that in the mornings I would be sent to 'suitable' clients suitable in the sense that the plant was in fair condition. After writing up the report I would return to the office for debriefing.

The first job was in Rochdale, the address merely the mill name, and the boiler turned out to be a 1910 vintage Lancashire type, I later found out that 1910 was fairly modern in terms of mill boilers. The journey was by train and bus, fortunately in those days busmen had an encyclopaedic knowledge of industry in the area and how to get there. I was dropped off within yards of the mill gate. Walking towards the mill, uppermost in my mind was the importance of giving a good impression, that of an experienced surveyor on a routine job, anything but a rookie.

The receptionist phoned the mill engineer who arranged for the boiler fireman to conduct me to his office. The fireman promptly announced me as yet another trainee of uncertain origin, probably Cockney or Liverpudlian so much for my nonchalance!

I found out several months later that this mill was favoured by my mentor for trainees on account of the very understanding engineer and the good condition of the plant.

Despite three hours of most thorough examination no deterioration since the previous inspection was found. I had hoped to find that one obscure unheard of defect that would have resulted in a catastrophic explosion but for my diligence – no such luck!

In addition to the examination, trainees were required to produce a detailed drawing of every boiler inspected, showing all the dimensions and previous repairs so that the safe working pressure could be calculated.

With this in mind I re-entered the boiler and started measuring up the various parts, a lengthy and difficult task with such old plant. After about an hour the fireman appeared at the manhole opening waving a photostat sheet which turned out to be a copy of the maker's original drawing, far better than anything I could produce.

With that I packed up and went home to make a fair copy for the office session the next day. It appeared the mill engineer had a wad of photostats for the many trainees who came, but only distributed them after we had sweated awhile in the hot boiler measuring up!

Despite the immaculate drawing I presented to my office mentor it only received a nod for approval. My fellow trainee who had been to another works sweated for about three hours, made his drawing with no assistance and received the same sized nod, life is just so unfair!

The remaining weeks spent at different premises on many types of pressure plant were largely uneventful but nevertheless interesting.

Further training was back in head office in the 'Engine Department' which was noticeably more relaxed than the previous department. It covered nearly every type of mechanical plant from large steam engines to plastic extruders, basically any plant that in the event of breakdown would cause substantial loss to the owners in repair costs or loss of profits.

The inspection regime was intended to if possible

forestall breakdowns which would be advantageous both to the clients and ourselves. Due to the wide differences in plant types there was little point in field trips, except for large stationary steam engines found in mills.

Here, as part of the service, efficiency checks were carried out by ourselves at regular intervals by attaching an instrument known as an 'indicator' to the engine.

The instrument consisted of a rotatable drum about two inches in diameter over which a length of graph paper was secured. The drum was rotated back and forth in unison with the engine crosshead by a system of cords and pulleys, replicating the movement and relative position of the engine piston.

A small piston within the instrument was connected by piping to the engine cylinder, the varying pressure within the engine cylinder was replicated at the instrument piston which was linked to a stylus tracing out the varying piston positions and steam pressures onto the graph paper. By these means a graph was produced, the horizontal axis representing the piston position and the vertical axis the cylinder pressures. The completed graph resembled the shape of a boomerang, from which could be extrapolated such matters as the horse power developed, valve settings and general efficiency of the engine.

When most steam ships were propelled by triple expansion engines, indication was part of the daily routine for the second engineer so that maximum efficiency could be maintained. I went to sea when the steam turbine was the main power source, hence I was new to the black art!

The modus operandi for indicating started at the close of the previous day's working and with the engine stopped, the client's engine tenter or minder would prepare the engine for the test by fixing the cord and pulley system, ending with a hook for attachment to the indicator instrument.

The first engine I indicated was a two thousand horse power cross compound of about 1880 vintage, a beautiful engine with the bedplate painted dark green, the cylinder covers and motion work highly polished, all brasswork similarly polished, the flywheel of some thirty tons rotating at a stately ninety revolutions per minute.

The engine house, green tiled to a moulded mahogany dado rail, above which were several courses of white tiles before cream paintwork gave way to an ornate wooden ceiling – a magnificent sight.

The engine tenter or attendant was nowhere to be seen and I started to attach the indicator to the engine. Almost immediately the attendant appeared and rained fire and brimstone upon me as according to him I had left fingermarks on his pristine paintwork! He informed me in no uncertain terms that on future visits, on no account was I to touch his engine. He had a point as, apart from the fingermarks, I really should not have touched the engine without permission and I profusely apologised.

Later he brewed up for me and we had a chat, it turned out that he had had about thirty years at sea as a greaser, hence his enormous pride in the appearance of the engine and its well-being. We shared seagoing experiences for the next hour and parted good friends. Sadly, a year later the engine was scrapped and the engine house demolished. I only hope he was not there to witness the destruction.

The next stage of training was with the 'Special service department' or 'Consultative department'.

I eventually did quite a lot of work for this department which was later simply called 'New construction' neither of the previous titles were to my liking – they sounded slightly ostentatious!

Engineering insurance firms have a vast experience of technical mishaps and failures and this knowledge is put to good use in advising manufacturers when it comes to the design of new plant. Typically a manufacturer would contract us at the design stage to look at the drawings and pass constructive comment, thereafter we would carry out stage by stage inspections to ensure proper workmanship.

The training comprised accompanying surveyors as they carried out the stage by stage inspections. A typical vessel under construction would be inspected at the raw material stage, checking the plates and confirming the physical and chemical test certificates. The next stage might be when the shell and ends are assembled for welding, checking the correct weld preparation, post welding scrutiny of the radiographs and so on - fascinating work!

Finally, the insurance side of the job. Not every client takes or requires insurance, the training was about definitions and wording. The last words in the training was about ethics, perhaps not a word readily associated with an insurance company!

Much to my delight the firm put this at the very top of all their activities and under no circumstances was this to be compromised. Whenever a mishap occurred we would be expected to be at the client's premises within an hour and on the spot admit or decline liability. If in doubt we were to give favour to the client and in all cases assist the client to expedite repairs.

During my subsequent career I cannot remember ever visiting a distressed client later than two hours from the mishap, nor was there ever a dispute over liability.

There are, of course, several other engineering departments within the firm, principally covering lift and crane equipment, electrical and gasholder, all carrying out statutory inspections with similar training schemes existing within each.

***

3. The History of the Engineering Insurance Industry

BOILER explosions and serious engine mishaps were a regular feature of the first part of the industrial revolution, incurring considerable loss of life, injury and commercial loss.

The only government involvement was by means of the coroners' courts, which resulted in a simple investigation and verdicts strongly suggesting that acts of God were to blame.

Alarmed at the frequent explosions and mishaps, a group of leading industrialists formed an association with the object of finding causes and remedies.

Thus the Manchester Steam Users Association was formed in 1854. Members included leading engineers, the then William Fairburn, Sir Joseph Whitworth of Whitworth thread fame and also Henry Houldsworth, a leading cotton spinner.

At that time the industrial revolution was well under way, steam power was the new rapidly evolving technology. Like many new ventures in the field of power production it carried risks and the object of the association was to minimise those risks.

A well respected chief engineer was appointed and he in turn recruited several inspecting engineers. It did not take long to find the main causes – poor design, materials, workmanship, maintenance and operation.

Members of the association benefited from regular inspections of their plant, enjoying a very considerable reduction in mishaps and explosions.

In 1862 a total of 458 boilers were inspected and none under this inspection regime exploded. However the bulk of mill and factory owners were not members and the general public as well as employees were increasingly alarmed at the continuing rate of injury and loss of life.

At that time Parliament was not lacking in wealthy industrialists who were hesitant to enact legislation that would cost them money and perhaps curtail their business interests.

Accordingly a clever piece of appeasement legislation was passed whereby in the event of an explosion the relevant government department was to be informed and a formal investigation would take place.

In the event that the owners were found negligent then court proceedings would follow. The Act was known as 'The Boiler Explosions Act 1882' it was intended to satisfy the general public and the factory owners.

However, after a few years its effectiveness was questioned by an MP who asked the House how many mishaps had been investigated. There had been some three hundred but when he asked how many prosecutions had followed, the answer was just one!

A further act was passed in 1890 and from then to this present day further comprehensive legislation has been introduced, not only for steam boilers but steam and air receivers, lifts, cranes and many other types of plant deemed to be potentially dangerous and requiring periodic thorough inspection.

In 1858 a rift occurred in the association with several members of the opinion that it would be advantageous to offer insurance to cover the particular risks connected with engineering plant. They broke away to form the first engineering insurance company, the predecessor of British Engine Insurance, shortly afterwards all the firms offered insurance.

The engineering insurance industry grew rapidly at this time and several new firms were started and the scope of inspections increased to cover many types of plant. In general, legislation followed the path already beaten by these firms.

Originally inspection was on a voluntary basis but slowly over the years legislation was introduced requiring compulsory inspection of those types of plant deemed potentially dangerous.

There are therefore two aspects of inspection, firstly to satisfy legislative requirements and secondly to prevent breakdown or failure. The concept of private insurance companies carrying out inspections on behalf of and paid for by their clients in accordance with legislation has sound historical roots and works surprisingly well. The motivation to carry out first class inspections is primarily a moral one backed up by law, placing stringent requirements on individual inspectors and their employing companies. Sound inspections greatly reduce the incidence of claims, a worthy cause in every insurer's mind.

In the rare case where a serious dispute occurs with an inspector or insurer the client can in effect sack them and turn elsewhere for inspections.

The engineering staff are divided amongst several broad disciplines, mine happened to be pressure vessels and mechanical plant.

The Engineer-Surveyor has an interesting role in our industrial society, he is employed by a commercial firm to carry out safety inspections at his firm's clients on a contractual basis. His is a role of advice and recommendation, his standards are to be found in codes of practice and the legislation and the advised methods of repair are usually based on the decades of experience accumulated by his employing company.

Since the promotion of plant safety is his major concern it might reasonably be asked how does he achieve this with mere 'advice and recommendation' in the tough commercial world which is not always lacking in the concept of profit before safety.

The answer rests in the relevant legislation, the surveyor is defined in law and named as the competent person, his reports have to be on prescribed forms and on serious safety issues a copy has to be forwarded to the government enforcing agency who, if necessary, will take legal action to enforce the requirements detailed on those forms.

Surveyors work on their own, visit clients, inspect plant, make reports, make decisions and talk issues through with clients; technical competence is the basic requirement.

The system works well with very few disputes or recourse to legal enforcement and surveyors generally have a good rapport with clients. In my experience I never had any problem with technically competent clients. I have however had a few interesting differences with, say, the other ranks, but more of that later.

***

4. Spreading oil upon the waters

THERE was nothing unusual about the plant, consisting of a large fuel oil storage tank supplying a package steam boiler, all made to a good standard and well maintained. The boiler was due for inspection on an Easter Sunday and was being prepared on the Friday and Saturday.

I was surprised to be called out on the Saturday to investigate the cause of the total loss of all the fuel tank contents causing severe pollution to an adjacent watercourse.

The supply pipe from the tank to the boiler was about 40' in length, fitted with a stop valve at the tank and a second valve outside the bund or enclosing wall. The pipe was wrapped with an electrical heating element, rather like an electric blanket, to ensure that the thick fuel oil was free flowing.

At the boiler front a spring operated fire shut off valve was fitted adjacent to the boiler. This followed the usual design of a fire valve kept open by tension in a thin stranded steel wire anchored at the remote end, in places likely to be vulnerable to fire, such as the boiler front, fusible links are fitted. In the event of fire the links part and the fire valve shuts, closing off the fuel supply.

The stop valve adjacent to the bund wall was found to be completely shattered, allowing the tank contents to freely escape. The valve was of good quality and designed to resist at least one hundred times the pressure imposed by the head of oil in the tank

It was obvious that an enormous pressure must have arisen to shatter the valve and cause the oil loss. Such pressure could only have been generated by thermal expansion of the oil within a confined space. Investigation revealed that the day before the boiler was due for cleaning, the works' engineer had shut the second tank valve as a precautionary measure but had not turned off the electrical trace heating. On arrival the boiler cleaners removed the fusible links and the associated control cable to facilitate cleaning. This act caused the oil shut off valve to close and render the piping a closed system, several hours later with the trace heating on, thermal expansion inevitably fractured the valve.

This was a good example of the usual requirement for two mistakes to cause a mishap. Had just the fire valve been shut and the heating elements left on no mishap would have occurred. The mistake was to shut both valves with the heating still on.

Many mishaps occur during unusual circumstances, holidays, shut downs, change of staff, early hours of the morning – written procedures can be an answer to this.

***

5. Embarrassment

AN unusual event once occurred at one boiler house which had the potential to cause me some embarrassment.

I had examined the boiler on Easter Monday and found no serious defects or repairs required, I left the premises and the works staff proceeded to prepare the boiler for work the next day.

First thing on the Tuesday the works' engineer phoned me to say that they were unable to use the boiler as it was leaking from nearly every tube like the proverbial sieve and could I come down post haste and explain why I had passed the boiler.

It was definitely not the kind of comment to promote one's view of being a competent person. (Incidentally, the term 'Competent person' is not mine but appears in the Factories Act and defines those deemed suitable to carry out pressure vessel inspections).

Anyhow, I returned to the source of embarrassment with some speed, with the intention of personal damage limitation.

When I arrived the boiler was surrounded by the works' technical staff, each expressing an opinion of the problem - and me.

I managed unobtrusively to get through the gathering and arrive at the boiler front. The smokebox doors were open and sure enough, water was dribbling out of most of the tube ends.

The situation called for some rapid investigation. I recalled that at the inspection on the previous day the boiler was generally damp which was a little unusual. This was explained by the fact that the flue or combustion side had been cleaned by water jet washing when usually this type of boiler is cleaned by brushing in a dry state.

Very often boiler tubes become slightly bowed in service, due to the fact that they are held rigid at both ends by the tube plates and any expansion of the tubes by the effects of heat can cause them to deform in the only way possible – bowing. The bowing can be upwards, downwards or sideways and is not generally detrimental. It occurred to me that with water washing, those tubes bowed downwards could hold a shallow amount of water at mid-length tapering to nothing at the tube ends.

When the waterside of the boiler was being filled with water it appeared that the additional weight on the foundations caused the boiler to dip at the front end by perhaps a couple of millimetres, sufficient for any water within the tubes to start dribbling out down the tubeplate.

This was certainly leakage, but not caused by defect. The presence of this water was demonstrated to the clients by having a portable light shone at the rear tube ends and observing the trapped water from the front tube ends.

So it all turned out to be much ado about nothing, a very good day for all concerned!

***

6. Reactors and reactions

DURING holiday periods surveyors often help out in adjacent districts and carry out inspections at unfamiliar premises. On one such occasion I was called out to a large chemical works to inspect a reaction vessel about 10' diameter and some 15' high.

As with most inspections the first port of call was to firm's chief engineer to discuss the inspection procedure and also to browse through any previous inspection reports as history is all important in industrial surveys.

We were interrupted in our deliberations by a request that the firm's safety officer would like to see me before any inspection was carried out. We duly walked round to his office and were invited to take seats whilst the safety officer completed a phone call.

Now first impressions are important to me so whilst half listening to the tone of the phone conversation my eyes were scanning the multitude of framed certificates adorning the office walls - all exalting the owner - and the piles of safety information hand outs on various bits of furniture.

My first impressions were reassuring and as usual I could not wait to see whether subsequent events would validate those thoughts. I didn't have to wait long.

As soon as he put the phone down he went into high pressure lecture mode on the imperative that I rigidly adhered to the firm's safety code - hard hat, bomb proof torch, antistatic shoes and so on at the same time thrusting into my hand a great safety manual the size of the Old and New Testaments combined and requiring the same amount of study time.

I'd only come to inspect a reaction vessel, certainly not to spend the summer studying his set book and his whole manner displayed insecurity, lack of confidence and rigid lip service to his responsibilities. I was inwardly fuming at his hubris but hopefully did not make it too obvious.

Once outside the office, the chief engineer, a most convivial and courteous man profusely apologised for his colleague's behaviour explaining that I was certainly not the first to be treated so badly. He also said that his colleague seldom left his office to look round the plant but preferred to stay put and bombard visitors and staff with hand outs and manuals thereby thinking he had absolved himself of all safety responsibility.

With this promising start to the day I walked over to the plant with an assistant to escort me to the reactor, carry my case, permit to work and most vital of all the hefty manual.

The vessel to be inspected was on the third floor of an open steelwork building reached by chequer plate steps. The assistant bounded up to the first level and expected me to follow.

I explained that although my job was simply to examine the vessel, I had also been very forcibly informed by the safety officer that all my actions were to be in keeping with the firm's safety policy, hence I had to ensure that I did not put myself at any risk. The handrails and every single step were most carefully examined, all were filthy and slippery and most of the steps had either bolts missing or slack.

My notebook worked overtime that day. The same attention was given to the next two floors, several pages of defects were logged. By then we had reached the real object of the visit, the reactor. This protruded through the third floor by about six feet and had a hemispherical top with a standard manhole. In normal circumstances I would have climbed up and entered through the manhole, but on this occasion with the lecture ringing in my ears and the near presence of the manual I had no choice but request a suitable ladder.

The assistant disappeared and about twenty minutes later brought a six rung ladder and also a long rope ladder. I did not have to carefully examine either, the rigid ladder had alternate rungs missing and both styles were split. The rope ladder appeared sound but was totally unsuitable for entry to the vessel. I had no intention of swinging around like a trapeze artist in a vessel containing lethal blades and baffles even though they would be locked and stationary.

For me, this was the last straw and the assistant was told that the site was far too dangerous and lacking in elementary safety for me to continue and therefore we should return to the safety officer, not forgetting to bring the ladder.

Meanwhile I added a few more notes and emotionally prepared myself for a good set too with the man whose profession was meant to be safety. The hour of his nemesis was nigh.

I first ensured a larger audience than that attending my earlier humiliation and then in courteous manner explained the short comings of safety with respect to access to the plant over which he claimed to hold sway. After careful recitation of my extensive notebook jottings and due delivery of the rotten ladder I informed him of my imminent departure. My final words being the effect that when he could ensure safe access to the reactor I would be pleased to return and carry out the inspection.

After about a week I was informed that all was ready and I carried out the inspection. The access was perfect all the walkways and railings were spotless and every nut and bolt tightened. The safety officer, however, was nowhere to be seen as he had wisely decided to take a rain check that day!

***

7. No fire without draught

WITH the demise of mill steam engines the associated boiler plant was kept for supplying heating steam during the winter months only and during summer months inspections and repairs were carried out.

One October I got a call from a mill engineer to say that he had made several attempts to fire up the coal fired boiler but without success and could I visit and help.

The boiler was a standard Lancashire type, two furnaces but only one in use, coal fired by hand, no economiser, horizontal brick flue about thirty feet long leading to the chimney.

On arrival the engineer explained that the fire beater (Lancashire for boiler attendant) had as on previous years laid firewood and coal in the furnace but on this occasion when lit smoke and flame merely puffed out at the boiler front rather than travel down the furnace.

To my mind this could only be caused by insufficient draught, the dampers were checked and found fully open, I found a door into the flue and walked to the chimney base, there was not the faintest draught up the chimney. I recalled the summer had been cold and exceptionally wet, this had undoubtedly cooled the chimney brickwork right down to ambient temperature, hence the lack of draught.

The solution I suggested was to build a large bonfire of timber in the chimney base to create draught, then shut the flue door and ignite the furnace fuel, this was promptly done and in next to no time the boiler was functioning normally.

With mills changing from steam power to electric drives the older steam experienced engineers were upon retirement replaced quite rightly with electrically experienced engineers. This was the case at this mill and what would be obvious to me would not be so to the current engineer [pardon the pun].

***

8. Claims and confidentiality

BOILERS occasionally suffer catastrophic damage through shortage of water, usually caused by the lack of maintenance of the automatic controls. Such damage invariably results in an expensive insurance claim.

The boiler in this case was a small package type and had recently been examined by me and nothing much untoward was found, I was therefore surprised to be called upon to investigate a serious mishap.

Considerable damage had occurred to the boiler and on examination it was found that nearly all the tubes were distorted and many were leaking at the front end plate attachments.

There was very little scale in the boiler and it was obvious that it had run short of water, I requested that the float chambers of the automatic controls be dismantled for close inspection but nothing was found to be amiss.

The elderly boiler attendant was questioned about the events leading up to the mishap, and he explained that he regularly tested the controls and that they always appeared to be in order. I was not too impressed with his competence but had no good reason to take the matter further.

Liability was admitted and repairs authorised, these involved a complete retubing of the boiler and final check by hydraulic test – not cheap!

Several days later, I called to witness the testing of the safety valves and automatic controls. I asked the attendant to raise the boiler pressure and also to demonstrate how he tested the controls. Sadly, he hadn't the slightest idea about either and worse still he left the devices in the wrong position – set up for another major mishap.

After correcting his mistakes I decided to have a quiet word with the works manager.

I explained my concerns and he appeared delighted that my thoughts were similar to his own. He explained that the man had been appointed to the job not on merit but because he was related to the managing director and apart from this fact would never have been given the job.

This was not going to be an easy situation to remedy as clearly the cause of the previous mishap and the likelihood of a repeat was due to the incompetence of the attendant, but to put this in a report to the clients could be difficult. My remit was technical not personnel but to ignore the situation would almost certainly invite another mishap, perhaps more serious.

In the end I decided to write an internal memo to my head office explaining the situation and suggesting that our commercial department meet the client's brokers where the matter could be discussed without committal to paper.

I was not present at the subsequent meeting but apparently during the discussion the broker requested sight of my letter to clarify a point and forgot to return it, he then a few days later conveniently showed it to the managing director who apparently went berserk at my carefully written letter. Not long after I was banned from attending the premises. I comforted myself with the thought that this was democracy at work - if you don't like the inspector get another one, I just hoped he was a bigger so and so than me!

***

9. Repairing the House of Cards

'MECHANICAL breakdown of calender...loss of profits policy in force... attend asap! This was the terse phone message from my head office one day.

Whilst driving to the textile mill where the mishap occurred I was racking my brains as to what a calender actually was, certainly don't find them on ships!

Loss of profits was explained to me during my induction, as I understood it, substantial sums of money were paid out by us to cover lost profits due to breakdown of a specific machine, the payout mounting day by day until the machine was back in production - hence the asap.

Having met the manager and chief engineer I was shown to what remained of the machine.

At this point I must explain how the machine would have looked in its prime, which coincidently was 1890 circa.

It consisted of two substantial cast iron box section vertical columns, each about 12' high and about 8' apart ,between the columns would be about six steel rollers each being about 15" in diameter, the rollers arranged one above the other. The rollers were either rubber or composition covered and rotated by a chain drive through a gearbox and electric motor.

The purpose of the machine was to impart a suitable surface finish to various kinds of cloth, this was achieved by passing the cloth between the rollers which themselves were forced together by considerable pressure.

The covered rollers were of slightly different diameters thus imparting a rubbing action to the passing cloth and determined its finish. An interesting feature of the machine was an ingenious device to separate the rollers when it became necessary to rethread cloth between them. The device comprised a vertical shaft in each column upon which were machined portions of thread which corresponded with threaded holes in the cast iron bearing housings for each roller. The shafts were coupled by a chain driven via a small motor and gearbox. The bearings and hence the rollers could be forced together or separated by rotation of the shafts.

The scene before me resembled a collapsed house of cards, the two columns remained intact, but the rollers, bearings, shafts and gearbox parts were strewn around the column bases, broken and bent.

It appeared that the bearing housings for the top roller had either come adrift or had broken in some way. This enabled the roller to drop onto the next roller and so on, a cascade effect until the whole machine was wrecked.

In situations like this there is a rightful expectation on the part of the clients that we as insurers will come up with a solution. I was also aware that hour by hour we were paying out for the commercial losses incurred.

With the machine nearly a hundred years old, the maker out of business, no quick trip to a spares department and certainly no drawings available, the prospects of a quick repair were poor, especially as many of the broken parts weighed several tons. There was one redeeming feature however. Really old machinery was built to last, the proportions of the various parts are usually fairly generous and provide the possibility of repair. The alternative was to buy a new or second hand machine, but this might involve months even to find a machine let alone transport and installation.

Whilst mulling over these options I arranged for the works' staff to sort out the wreckage so that each piece could be carefully examined and assessed.

This exercise took several days, the most difficult task was to sort out the pile of broken cast iron, rather like a huge jigsaw. The various bits were loose assembled on the floor to give some idea of the original components.

The working surfaces of the covered rolls were badly scuffed and abraded and the two screwed shafts were bent and beyond repair.

So, with the full picture before us, it had to be repair with all possible haste. Various specialist firms were called in with permission to employ as many staff as could actually work on the repairs, round the clock working and as much overtime as required - could have been a main engine breakdown at sea!

The covered rollers were transported to a local firm specialising in this type of work and restored to as new condition, the screwed shafts were difficult to replicate as the bent originals were also badly worn and a judgement had to be made as to some of the original dimensions, but again a local firm made an excellent job.

The broken castings presented a real problem, some were beyond repair as the pieces were much too small to restore to working order. However the first port of call for broken castings has to be a firm using the metal stitching technique, this involves rejoining adjacent broken pieces by the insertion of precision metal keys.

The process requires the parts to be clamped, then machined in situ, the keys then driven in, the effect being to tightly pull the fractured surfaces together. This is work that requires careful design and selection of the sites for keys and very accurate workmanship. The result is a perfect job and I have seen may such repairs, none have failed.

The firm employed had a foundry, an essential department for reproducing parts damaged beyond the possibility of stitching. Most of the broken castings were successfully repaired by the stitching process, others replaced by new castings and then machined to the original dimensions.

Finally the rebuilding of the machine commenced and all was going well and within a few days it was ready to run except for the fitting of the screwed shafts.

But we quite unexpectedly met a major setback.

For correct operation the threaded bearing blocks had to be in precise synchronism with the threads upon the shafts, we could find no marks or other clues on the original shafts to ensure correct positioning.

Had the machine been much smaller then a little trial and error could have been used to get the correct positions, but with most of the parts being extremely heavy, this was clearly impossible. Despite all our efforts we had reached dead end and still the money was haemorrhaging away each day. All sorts of theories and ideas were aired but to no avail.

But suddenly we found gold! An onlooker to the repairs remembered an event about thirty years previous when apparently the machine had undergone a complete overhaul and our onlooker remembered the name of the foreman in charge of the work. Our commercial staff started to track this man down, apparently he had retired and gone abroad and was finally found to be living in South Africa. With sufficient financial inducement by my firm he was persuaded to take a week's holiday (?!) back in his old home town and invited to visit his former haunts!

He had lost none of his skills and soon organised the works' staff in assembling the shafts, within a day he had the machine back into full production.

Cards is largely a game of luck and in the restoration of this house our luck certainly played the major part.

***

10 Industrial Central Heating Boilers

MANY large central heating boilers of the type used in commercial premises and factories are insured for such events as explosion, cracking and joint leakage.

Strictly speaking describing such equipment as a boiler is a misnomer as they merely heat water and never actually boil it and generate steam, but most people recognise the usual description.

Some of these boilers were of the cast iron sectional type comprising of from four to about fifteen sections, mainly coke fired or converted to oil firing, each section was of cast iron and hollow, through which water was circulated. The general shape was of an inverted letter 'U', each section was provided with ports to enable coupling to adjacent sections and a pathway for the water, the whole assembly being held together by longitudinal tie rods.

The combustion chamber being formed within the space bounded by the legs of the inverted 'U'. The purpose of inspection was to forewarn clients of impending problems so that remedial action could be taken and save future breakdowns.

One of the modes of failure is due to condensation of the flue gases, particularly whilst the boiler is warming up and the water is relatively cold. The sulphurous residues liberated from the combustion process combine with the droplets of condensation to form acid which runs down the sections and causes corrosion, even when the boiler is not working. During summer any residues left through lack of cleaning attract more moisture and the corrosion is continued. If left undetected and unchecked the sections - originally fairly thick - are corroded to be paper thin and ultimately burst causing a major breakdown.

This process is well known to engineer surveyors

and its extent can be mapped by careful hammer testing, the remaining thickness of a section can be fairly accurately ascertained, a moderate hammer blow on a thick section gives quite a different sound from that on a thin section.

From memory the particular boiler in question had over twelve sections and was in the basement of a large factory. Having met the manager he conducted me to the boiler and returned to his office.

The boiler had been thoroughly cleaned and a pile of ash and broken refractory brickwork removed during cleaning was piled about 2' from the open boiler front. Hammer testing revealed that the second and third sections from the front were reduced to paper thin at the lower sides, the usual site for such problems.

As the boiler was less than ten years old and the thin sections were at the front end, I reckoned that replacement of these two sections would be relatively easy and inexpensive.

I duly returned to the manager's office and found him with several of his staff, my findings were quickly described together with the advice that the two sections on the point of failure should be renewed if the boiler was to remain reliable for the coming heating season. Possibly bolstered by the audience of his fawning staff he waded into me, making it very plain that neither I nor anybody else could possibly determine that the sections were thin and he would like some proof of this ridiculous assertion.

Unknown to him, the display of such blatant pomposity at my expense triggered my brain to consider the possibilities of taking some form of deflationary action.

At my suggestion we made our way back to the boiler for the proof he sought. Having a good idea of the direction this proof would take, I advised him to enter the combustion chamber backwards so that from the outside I could face him and demonstrate the art of hammer testing.

Bearing in mind he had a duster-coat over his pin striped suit and was crouched in the confined space, I struck the second section with abundant zeal! Not surprisingly the hammer head went clean through the thin metal, punching a hole about 1½" diameter through which an enormous volume of water shot out, propelled by the static head of about 40'. The water shot from left to right across the boiler, quickly flooding the boiler house floor, the ash pile soon disappeared with the rising tide. A few bricks remained visible onto which I leapt, marooned, but relatively dry.

The proof seeker was trapped within the boiler, not daring to escape by passing through the intense water jet, but nevertheless still getting soaked by the swirling spray. Despite my encouraging words to make the leap of faith out of the boiler he remained huddled up at the back, in due course the flow subsided and our man bent double crept out.

Unfortunately he tripped on some under water obstruction and fell headlong onto the flooded floor. Momentarily prostrate and semi submerged on the flooded floor, his appearance suggested an oversized frog.

I had my problems as well, perched on a brick trying to keep balance is no fun when at the same time attempting to subdue the possibility of explosive uncontrolled laughter, I was petrified in the knowledge that like an encounter with the diarrhoea, ultimately I would lose control and act totally out of character! To my immense relief he stirred and slowly arose like homo erectus from the primal soup and paddled across the boiler-house floor to dry land.

Glistening with ooze, suitably blackened and saturated, he made his way up the steps more by feel than sight to his office where his informed staff gave the hero of No.2 Boiler-house a tumultuous welcome, hopefully followed by a full towelling down.

Unfortunately, another esteemed client at the far side of town awaited my service, so I could neither take the salute nor hand out the medal! As for myself, I had followed at a discrete distance, expecting a counter attack. Intuition told me that a state of tension might well exist between us, ever the tactful diplomat I tried to defuse the situation by opening a conversation on the merits of cast iron, but between the coughs and gurgles his preferred subject appeared to be in some way connected with my ancestry.

Having fled to my car the suppressed ecstasy took over, any passing policeman would have been within his rights to arrest me on the grounds of drug or solvent abuse. I soon came down to earth as I pondered whether my chosen career might be heading for derailment, it seemed inconceivable that the deflated one would not take some form of retaliatory action.

With these thoughts high in my mind I planned my defence, fortified with lashings of humble pie, I would plead before my head office interrogator that my rottweiler reputation was completely unjustified, I was in reality any client's poodle, always at their beck and call, day or night. Even with this complaint. I had gone out of my way to satisfy our esteemed client's curiosity, the mishap was just another Act of God and we of all people as insurers were well used to that.

On the way home I called at WH Smith's and scanned the shipping papers and was relieved to find plenty of jobs on offer. In fact my old firm 'Manchester Liners' was advertising in several of the papers for marine engineers!

Surprisingly, the telegram requiring my presence at head office never came, however the anticipation of its possible arrival did raise my exceptionally low stress level and I resolved that in future I would be much more subtle whenever rebalancing action was required.

Some eleven months and twenty nine days later, just within the period for the next contractual inspection, I, reformed and contrite, revisited the works and was delighted to find that my advice had been taken, two new sections had been fitted and that the manager had found pastures anew. I was however a little disappointed at being unable to resume my conversation on cast iron!

***

11. Plates and Ladders

A FUEL tank situated in a factory yard required inspection and as is the usual practice it was protected by a bund wall to prevent spillage or leakage into surrounding land.

This particular part of the yard was not surfaced but was covered with long grass. The wall was about 10' high and since I needed to reach the tank a ladder was requested and duly leaned against the wall with one of the factory staff steadying it for me.

I had almost reached the top when without warning I and the ladder dropped into a deep hole that appeared at the foot of the ladder, ending with my head at ground level. Instinctively scrambling upwards from the abyss, I reached the surrounding solid ground. The ladder had disappeared and my shocked assistant was speechless.

Investigation showed that beneath the grass was a 6" layer of soil under which was a large steel manhole cover giving access to a 6' diameter bricked culvert through which a considerable torrent of water was flowing. The cover and its frame must have been severely corroded and the concentrated weight of the ladder and myself caused it to collapse, the manhole cover and the ladder were swept away down the culvert. 'Don't step on the grass' had a new meaning for me!

However, there was no way this event could have been foreseen and no doubt the insurers' favourite phrase 'An act of God' came into play which was also applicable to my survival!

***

12. Problem Piping

STEAM piping is to be found in almost every factory utilising steam, connecting boilers to steam-using process machinery and generally causes few problems.

It therefore came as a surprise when a new extensive piping system suffered an explosion when a 5' length of the pipe split open.

Steam piping requires proper supports, adequate provision for expansion and contraction and also suitable arrangements for the removal of condensate.

The piping in question fulfilled this criteria and was in all other respects built to the highest standards, such as materials used and quality of the various welded and flanged joints.

The piping was at a large textile mill and replaced an ageing system. I was particularly interested, as over the previous six months I had been carrying out stage by stage inspections as the replacement pipework was erected and had signed off the installation as in good order only a few weeks before the incident.

The normal working pressure was 150 psi. and mainly consisted of between 6" and 8" diameter steel piping, the damaged portion was within a 60' horizontal length.

As in all such cases the damaged portion was cut out for close visual and metallurgical examination and nothing untoward was found with the material, it being sound and of good quality. Whenever good quality steam piping explodes and opens up in this fashion invariably the cause is water hammer.

The mechanism for this phenomena is well understood. Steam entering cold pipework condenses and if not drained away from horizontal lengths of piping, forms a pool of water over which the advancing steam passes and condenses even more rapidly. This leads to a partial vacuum at the remote end of the pipe, which rapidly accelerates the standing water in the same direction. The momentum of this water, if resisted by a closed valve, can generate an enormous force able to split pipework or fracture a closed valve. In these situations the actual steam pressure is usually quite low during the warming up of the piping; it is the velocity of the condensate that does the damage.

Well-designed pipework such as this installation makes provision for the automatic removal of condensate by means of devices known as steam traps. These are fitted at the low parts of the piping, detect the presence of condensate and discharge it to a suitable drain.

In this case the incident happened at two o'clock on a Monday morning. The works had been shut down for the weekend, steam was raised in the boilers on the Sunday evening and admitted to the piping early on the Monday. The boiler fireman was unaware of the incident as it occurred in a remote part of the works, the escaping steam severely damaged some newly installed machinery.

Having regard to the fact that water hammer was the probable cause of the incident, the condensate drainage arrangements in this particular length of piping were closely examined. The piping was protected by several steam traps and these were dismantled and found to be in good working order. The drainage piping for conducting the condensate away was checked for free passage and found clear.

Whilst investigations as to the cause of the mishap were taking place, that part of the factory was shut down. It was much too dangerous to permit normal working without having established the cause and put in place preventative measures.

It was decided to replace the damaged portion and this merely involved electrically butt-welding a new piece into position and then verifying the weld integrity by ultrasonics. Steam was admitted to the pipeline and some of the works staff and myself patrolled the area checking on the functioning of the traps and also listening for any signs of waterhammer.

The piping was brought up to full pressure and that part of the factory resumed production. With the piping at normal working temperature, water hammer should not occur even if faults were present. The situation was not ideal but it was at least safe whilst the piping was in use. The danger time was after each weekend shut down when steam was admitted to cold piping. To offset this, staff were employed at the crucial times to check during the warm up period, this procedure went on for several weeks and never once was water hammer detected.

My lucky break came due to the vigilance of a night watchman, employed by a security firm to patrol the factory over the weekends. In his log book he recorded the fact that whilst in one of the wash houses he had heard a cistern continuously filling but was unable to locate the exact source. The factory engineer picked up this comment and asked the plumber to investigate.

He found that the problem was not a cistern but a shower unit. The unit was unusual in that steam and mains water were mixed within a chamber to provide hot showering water. The steam and water supplies were protected by non-return valves so that water could not enter the steam piping nor could the reverse happen. The hissing noise the watchman had heard was due to the failure of the valve in the steam supply line permitting mains pressure water to enter the steam line.

Over the weekend shut down period this had flooded the length of piping in question, the volume of water present was too much for the traps to cope with and on admission of steam, water hammer was the inevitable result.

The faulty valve was examined, a ¼" diameter nut had slackened off because the manufacturer forgot to fit a split pin, the loose nut intermittently seized the mechanism.

This was an accident almost impossible to foresee and prevent, but all praise to the watchman. His vigilance and enthusiasm for the job certainly prevented a further, perhaps more serious, mishap.

**

13. Safe Landing

MANY mill engine houses share a similar design, with the engine being about 12' above ground level. Substantial foundations extend through a basement area in which are located the ancillary equipment such as boiler feed pumps, condenser and circulating pump.

The engine room floor is reached by external steel steps up to a platform, similar to a fire escape, beneath which are usually stone or brick steps down to the basement some 10' feet below ground level.

It was a Bank Holiday and part of the engine was adrift for examination, a routine job. On completion I stepped onto the outside platform preparing to leave, case in hand and waved a farewell to the mill engineer.

Immediately, the platform collapsed and fell through onto the stone steps below, where it shattered into innumerable pieces. Under gravity, I accelerated downwards at the engineers' favourite constant - 32' per second, per second!

The 'Act of God' followed, I managed to grab the lower hand rail that encircled the platform, the case containing instruments and papers tumbled down and joined the mass of broken cast iron on the cellar steps. The engineer rushed out and pulled me up and clear of the debris. He was amazed I had survived, expecting to see mincemeat on the cellar steps. He rescued the case with a pole as it was much too dangerous to walk on the wreckage.

Together we investigated the cause of collapse, which in fact was obvious, the cast iron platform had been supported on a substantial angle iron frame, which in its near hundred year life had corroded to paper thin in the hidden parts. To this day I am wary of steel supported platforms, always keeping a firm grip on the handrails.

At the time I had only been in the job a few weeks

and was living at home in Wallasey, I casually mentioned to my factory inspector father that I had had an interesting incident. He immediately wanted to know how, when and where, I refused to divulge this information which caused a temporary rift between us. He saw every dangerous incident as a breach of his beloved Factories Act and appeared to relish the possibility of a day in court – he was not one for a quiet word of persuasion.

There was an amusing sequel to this incident, but first some of the associated background. At that time the arrangements for inspections were initiated by my head office. When an inspection was due a letter was sent to the clients requesting a suitable date, clients would write back to the office with a date, the office would then write or send a telegram to the surveyor informing him of the required inspection date.

Pressure vessels are subject to statutory inspection periods and in the main these are adhered to by clients although they occasionally over-run, particularly with minor pieces of plant. Visiting factory inspectors make a point of checking the registers of inspections and in no uncertain terms bring into line those that have failed to keep up to date.

Such a client would inform my office and request an immediate inspection so as to avoid the possibility of a court appearance. Normally very few such requests would be made, but shortly after my mishap I received dozens of instructions from my office requiring me to inspect overdue plant in the same town.

The cause of this activity turned out to be that my father had flooded out the town with inspectors, instructed to find the factory with the damaged platform. In the course of this search they routinely checked the registers at all the sites visited, hence all the overdue inspections I had to deal with. They never found the damaged platform.

***

14. Take me to the cleaners

BEFORE any pressure vessel can be inspected it has to be prepared by thorough cleaning. In the case of steam boilers this can, depending on the size and type of boiler, take merely a few hours or in the case of a large power station boiler, many weeks.

The people employed for this work were generally either casual labour or the particular factory's staff seconded to the work. By far the worst type of plant to clean was the coal-fired Lancashire boiler to be found in almost every textile mill.

Every part of the fire or smoke side, every plate, flange and rivet had to be cleaned down to bare metal. Likewise at the waterside all scale and sludge deposits had to be removed, however, depending on the type and thickness of scale there was some slight scope for latitude.

The Lancashire boiler has extensive supporting and enclosing brickwork, a labyrinth of narrow flue ways below and to each side of the boiler shell. Coal is burnt in the two grates of the furnace tubes, the ensuing smoke and ash pass rearwards and down into the flues. Here the ash deposits to depths of up to about one foot. Within the boiler or waterside, scale is the problem. It deposits on all the heated surfaces and can be dangerous on the furnace tops less so at other parts. It can also be hard as rock or soft as chalk depending on the feed water quality and treatment.

This is the empire of the much-maligned scaler or boiler cleaner, a truly appalling job characterised by heat, dust, filth and claustrophobic working conditions. Heat is a major problem and despite several days of cooling down, the residual heat remaining in the brickwork and foundations - augmented by the radiated heat from working adjacent boilers - makes the work within the realms of the unbearable.

Suitably clad in overalls, complete with hood and goggles, equipped with short shovels and buckets these cleaners enter the 'hellhole'. The head room seldom exceeds two feet and the bottom foot is solid ash. Within minutes the disturbed dust reduces visibility to about one inch, hour after hour the ash is shovelled out and barrowed away. Meanwhile others are in the boiler itself, descaling with chipping hammers under the same working conditions but with noise for good measure.

Amazingly I have met some very proud and enthusiastic men in that job whose goal was perfection, they actually competed to present the best cleaned boiler. It made my job much easier and quicker and the perfection in cleaning was quite noticeable, considering at least thirty barrow loads of ash had to be removed, the amount remaining would not have filled a teacup.

Notwithstanding that the clients had a legal duty to have the boilers thoroughly cleaned for inspection, I made it a point to always seek out the cleaners and personally thank them. If unable to find them, then suitable appreciation would be put on the statutory report.

Happily with most boilers now gas or oil fired and the treatment of feed water maintained to a high standard, such hard labour intensive cleaning is seldom required, although coal fired power stations still require considerable cleaning effort.

***

15. Hydraulic Tests

THE hydraulic testing of boilers and pressure vessels is a long established method of detecting defects and latent weakness, particularly where contracted parts deny the inspector the opportunity of close examination.

Pressure vessels have to resist internal pressure and the best shape for this is either cylindrical or spherical. Most are of cylindrical construction and in order to form a closed vessel, end plates are required and these are usually flat or dished. Unfortunately flat plates are poor at resisting pressure and readily distort, unchecked progressive distortion - always curvaceous - can threaten the safety of the vessel. You wouldn't imagine a piece of 5/8" boiler plate could have such sex appeal!

Various devices are used to beef up the strength of flat plates, such as corrugations, stay rods, stay tubes, adjacent flanges and girder attachments but whatever method is used, the potential for excessive distortion remains, accentuated by corrosion, fracture, overheating and very occasionally design faults or poor workmanship.

A properly conducted hydraulic test presents the opportunity of detecting such possible weakness before it can develop into a catastrophe. Hydraulic pressure is applied in excess of the normal operating pressure of the vessel, the subsequent increased stress imposed exacerbates any latent weakness or leakage.

The procedure is safe, as water being virtually incompressible has negligible stored energy despite its pressure and any structural failure will be a fairly tame affair. In the event of say a fracture occurring, water trickles out and the applied pressure drops to zero.

There is no catastrophic explosion, which would occur if the pressurising medium was a compressible gas like air. For this reason it is vitally important that no air pockets remain in vessels to be tested as these would become a source of stored energy with the potential for danger in the event of a fracture occurring.

Prior to the test the vessel is thoroughly examined and locations of possible weakness noted. In general these would be the flat areas between strengthening systems, which if defective would permit excessive distortion at the increased pressure. On a Lancashire boiler for instance this would be on the front end plate between the gusset stay ends and the furnace attachments, or on a locomotive boiler on the firebox plates at the geometric centres of the many stay groups. Once marked the next step is to precisely measure the distances to known reference points on the boiler and make a note of the dimensions.

After filling the boiler and making sure that no entrapped air remains, it can be slowly pressurised by pump, at the same time the boiler is constantly checked for leakage and signs of distortion which if left could cause serious damage.

During this stage boilers and vessels occasionally emit squeaks and groans as the structures accommodate the rising stress levels. On reaching the test pressure required, the measurement points are revisited and checked, revealing hopefully a temporary distortion of only a millimetre or two. On slowly releasing the pressure the distortion should disappear completely, evidenced by a return to the initial dimensions. The all-important matter is that any pressure induced distortion is temporary, as permanent distortion would indicate weakness and merit further investigation.

Hydraulic tests are also useful for detecting leakage from seams, in the case of welded seams this is a serious defect requiring closely monitored repairs. With riveted seams this can in most cases be rectified by caulking.

However a history of persistent leakage may require investigation, hidden cracking around the rivet holes has the potential for explosion.

Hydraulic tests at industrial premises are fairly routine affairs, necessary if internal parts cannot be reached, or as a final test after repairs, but much more interesting are tests on restored antique boilers, particularly at the owner's premises.

If you ever get a complimentary ticket to attend a hydraulic test on a 1905 steam roller, take my advice and sell it. That is, unless you get turned on by nudity, then go early and get a front seat, for the boiler will be stripped naked, in its birthday suit so to speak.

However should the ticket secure entrance to a preserved railway then perhaps even if technicalities are not for you then the accompanying ceremonial will make the visit well worthwhile, particularly if you are aware of the various stages of the ceremony. The ticket date indicates the magnitude of the event.

If it is for a weekday it will be low key with just a few local devotees and no marquee, beer or cake. More than likely it will be running late as Professor Plum (club secretary) has forgotten to purchase the all-important 1 1/2" B.S coupling. Should the ticket show a weekend date you are in with a chance, this gives time for the gathering in of the many enthusiasts countrywide - no real disciple would miss this.

If nothing else you will be surrounded by enthusiasts all eager for a slice of the action. Don't forget the camera, posterity will be unforgiving if you fail them now. Anticipation floats on the air.

At this particular ceremony, just maybe, the long awaited cosmic bang will occur, the gathered enthusiasts who, despite their ten years labour of love, will not be too down hearted, the ascension of locomotive boiler No.267 to the heavenly engine shed, soon to be reunited with its partially restored chassis to become finally once more the 'Pride of Britain' class 7F calls for rejoicing. No puffer could have a better end.

I have picked the testing of loco boilers not because of any animosity I have towards them, but because no industrial boiler has quite the glamour or is loved so much by so many. Boilers at coal fired power stations have few lovers, celibate from birth to scrap man.

The ceremony is part religious, part secular and starts with the gathering, where the disciples mill around discussing anything from types of whistles to wheel arrangements. The trouble is they are just too knowledgeable, I have to avoid them like the plague. For me to be caught between two of them raises the spectre of acute embarrassment - I'm expected to be the fount of boiler wisdom but after a few minutes they could be talking Chinese for all I can understand. Fearful of opening my mouth in response to some obscure question, the only safe reply is a diagonal nod, neither 'Yes' nor 'No'...

I have requested and witnessed many such tests and noted the eager look of anticipation and fear on the faces of owners and enthusiasts, as the slowly rising pressure is perceived by some as possibly heralding disaster.

As previously mentioned this is the only time a boiler actually speaks its mind. The language is complex, you need knowledge and linguistic skill for interpretation - it is the language of protest. At the first creak or groan the assembled multitude nervously take a few steps back, not too far back though, fearful that flight may indicate that they are not seasoned hydraulicers.

Believe me, the really big bang and the rending of steel never happens. If you are extremely fortunate a small damp patch may occur around the odd tube expansion or perhaps an almost imperceptible drip is sighted from a seam.

Such momentous revealed defects have a magnetic attraction to the assembled congregation who barge and shove each other to lay hands on the imperfection, no doubt hoping in some way to effect a remedy for the defect.

I happen to be the inspector but there's little chance

of getting anywhere near the exciting discovery. When I eventually do get near, my stress level rises at being completely unable to locate the defect, knowing full well that their finding will be a minute speck of water less than that provided by a flea's urine sample. I find that the best way to locate the problem is to get one of the elders to walk round the 'Beloved' with me, he will be delighted to point out that which I have missed!

The synod then convene a meeting and seek my approval of the proposed remedy. I'm all for doing nothing but this is merely academic as during the endless discussion the minute trace of leakage has evaporated.

Pressure is then slowly lowered and with it the excitement and anticipation wanes, the handshakes and benediction follow - surely even TV is better than this, I am with you, a really good bang would make it all worthwhile!

The enthusiasts have a final card to play. As I am about to drive out of the main gate, after spending about an hour de-carbonising and degreasing myself, the side window receives a tap and a disciple appears with a question. Did I notice the crack at the lower left side of the foundation ring?

No, I didn't nor did he, the blighter just wants to cast doubt in my mind, such as to cause a re-examination. This can be quite disconcerting, should I have a second look or take the risk that this is his idea of a joke? My standard answer is foot to the floor, hopefully pelting him with gravel and surrounding him in rubber smoke. A glance in the mirror reveals surprise and disappointment!

Whilst tests on railway engines are now technically well conducted and run to plan, assisted by competent enthusiasts, the same cannot always be said of the tests of road locomotive boilers such as steam rollers or traction engines conducted by a sole enthusiast.

The opening scene betrays an epic struggle fought over the previous weeks, water everywhere, numerous lengths of garden hose, jointed and taped, stretched between kitchen sink and the 'Beloved', piles of discarded wrongly sized pipe fittings and amidst it all, our stressed out enthusiast struggling with a borrowed test pump.

With commendable optimism he pumps away whilst the pressure gauge needle flickers around zero endlessly, before capitulating to the fact that the pump is in a worse state than the 'Beloved'. Several days later with the pump resuscitated the test pressure is within a hair's breadth of achievement when the left hand mudhole cover starts uncontrollable weeping, perhaps distraught at its owner's travail. Out comes the omnipotent 36" Stilson wrench, the cover stud making its last painful stretch, snaps in two and the weeping becomes a deluge, once more the test is abandoned.

With touching tenderness the enthusiast cradles the deceased 3" x 4" mudhole cover, whilst the escaping water burbling and gurgling plays the sad lament. After a few more weeks our enthusiast requests my presence yet again, this time his efforts are crowned with success, unfortunately the test revealed a serious defect in the boiler which with any luck will keep him off my back for several months!

Tests on industrial boilers and vessels are pretty low key affairs compared to those on restored artefacts where there is always excitement and plenty to interest.

Finally, restoration is a long, expensive and arduous labour of love, the beautifully restored artefact is a testament to enthusiasm and persistence in the face of immense technical difficulties. None of the parts can be obtained as spares, if it's broken or worn it has to be repaired or remade with whatever tools are to hand. If specialised tools are required then they themselves have to be made first. The restoration movement, nearly all volunteers, has saved much of our industrial heritage and gives the rest of us much pleasure as we visit their sites of activity.

***

16. Economisers and Ethics

AS the name suggests, economisers are devices to save fuel by utilising the heat from otherwise waste flue gases from boilers to raise the temperature of feed water to be used in the associated boilers.

Nearly every textile mill boiler plant was equipped with such a device and they made an appreciable saving in fuel. There were two main manufacturers in this field, both used similar designs but varying slightly in detail. Depending on size, they consisted of batteries of cast iron pipes set in box section headers, each pipe being about 8' long and 3.1/2" bore. The water passed through the pipes and the hot gases circulated around the outside of the batteries.

This type of plant had a long and reliable life span and presented few problems from an inspection point of view. The main agents of deterioration were condensation on the outside of the tubes, caused by the low incoming water temperature and corrosion of the internal parts accentuated by soft water. Part of the inspection comprised of gauging all the tubes particularly at the lower ends where the condensation concentrated and was acidic due to the high sulphur content of the flue gases. Once below the calculated safe diameter, tubes had to be replaced.

At the internal side of the tubes the main problem was scale, too much could cause the tubes to fracture and since they were working at full boiler pressure this was dangerous. Furthermore the presence of scale reduces efficiency and impedes inspection.

The amount of scale was kept down by having the tubes periodically bored out by a power driven rescaling machine. Access to the tubes was obtained by removing a tapered plug from above each tube. Since the economiser had to be out of service for a week or so, this was a costly procedure.

After the training period at my firm's head office I was appointed to my district, let off the lead like a new puppy, to start inspections on my own, carefully watched from on high. Unfortunately within days I had crossed swords at the works of a major client. Looking back over the years it did me no harm but I could have handled it better.

The problem centred around an economiser as just described, I met the firm's engineer at the mill and had a preliminary look at the economiser. It appeared to be properly prepared for inspection, depending on previous history, it was usual for the clients to remove about 10% of the tapered plugs and this had been done. Whilst looking around, I glanced at the works' engineer who appeared ill-at-ease, nervy and like the proverbial cat on hot bricks.

I dismissed it from my mind and got down to the inspection. Shining my torch down a few of the tubes I was surprised by the complete lack of scale and eventually all 10% were checked and found to be free of scale. It occurred to me that perhaps the wool was being pulled over my eyes, perhaps these tubes had been bored out to give the impression that all the tubes were scale free.

That would certainly explain the engineer's demeanour. Accordingly, I requested that another fifty plugs be removed, but the engineer refused on the grounds that the original 10% were in response to an agreement that his firm had with mine.

This particular economiser had over 800 pipes and experience with similar pieces of equipment revealed it was generally only necessary to remove about 10% of the inspection plugs to form an opinion as to the state of the remaining pipes. Accordingly this loose agreement was made, but it did not override the law.

I must admit that in some way I relished the direction this was taking, it struck right at the heart of my interpretation of the relevant section of the Factories Act. As the inspector I set the pace and called the tune, not the client. It was my signature which went on the certificate not his and I took the responsibility not him. There was not the slightest possibility of me backing down.

Eventually after some polite persuasion he removed the plugs, nearly every tube was not only scaled but almost completely blocked. There was nothing else for it, all remaining plugs would have to be removed and more than likely every tube bored out. The engineer looked resigned to this possibility and mentioned that he would have to get authority from his chief engineer for this work to be done. I then left and awaited developments.

Within days a meeting was called at the works to meet the firm's chief engineer to which I duly attended and worse was to come. The meeting did not get off to a good start as almost every sentence spoken to me was prefixed 'Listen laddy'. I was thirty at the time, perhaps he would have preferred for me to sit on his knee and take instruction from granddad!

I don't take humiliation from clients and whilst he droned on about his elevated position, the power he could wield and the important people he 'golfed' with at my firm, my mind was working overtime as to how best to ensure a thorough debunking. He first acquainted me with the fact that his firm had over fifty boiler plants inspected by my firm and that he hadn't come to argue but to tell me that if I persisted in wanting all the tubes bored out he would remove all the business from my firm.

Things were getting a little serious, if in fact a client could wield this power, then to me there was no point in inspection and I might as well go back to sea. At that time there was a serious shortage of marine engineers, so I was not unduly alarmed. The meeting changed nothing, it was a matter of principle to me, I was being paid to make sure that plant was safe, to take no risks at all, we parted but not before he informed me of his intention to take the matter up with my firm.

Several days later a telegram arrived requesting my presence at my Head Office to discuss the matter with the chief engineer of boiler department - the client had given his side of the story and I was invited to give mine. It was not easy for my boss, there was the threat of losing business and maybe being a newcomer I could be seen as too zealous. In the end he decided to send a more senior engineer to look at the economiser and asked how I felt about that, I was quite pleased and willing to accept his verdict.

This engineer duly reported back that the economiser was in a dreadful condition and he could not possibly carry out an inspection until all the tubes were bored out - I was absolutely delighted with this!

In due course the mill engineer had all the tubes bored out and the inspection was successfully concluded. His chief engineer was not present, but I did send him a message that the economiser was just passable and that next year I hoped it would be a bit cleaner still...

***

17. Maintaining the Circle

MOST pressure vessels and boilers are of cylindrical construction as the circular cross section is the best shape to resist either internal pressure such as that which bears on the outer skin or shell, or external pressure pressing on such parts as the internal furnaces of the old Lancashire or indeed modern shell boilers.

Precise circularity throughout the life of the vessel is essential, any lack can be dangerous as demonstrated by the many explosions through this cause.

Many older boilers and pressure vessels have shell and furnace seams of the riveted lap type. Here the circular shape is constructed by rolling a flat plate to circular form and then overlapping and riveting together. The act of overlapping destroys true circular shape and a problem arises in shell seams because circularity is compromised at the lap. Internal pressure concentrates stress in the outer lap just beneath the internal lap edge.

Varying pressures and temperatures additionally tire or fatigue the steel shell at this very inaccessible place, generally beyond visual inspection. Eventually, the material fractures and an explosion follows.

The inspection solution evolved first by having blind or test holes drilled into the outer lap to almost the plate thickness and in line with the expected fracture, any fracture starting would reach the hole and leakage would be the witness. Next, instead of holes, slots were cut to the same depth across the expected line of fracture, the advantage being that any fracture that strayed from the usual position would be found. Both these methods could fail if the fractures occurred between the several test points. Finally, with the advent of ultrasonic testing the whole seam could be easily scanned - the perfect solution.

Larger and later riveted boilers do not have lap

joints. Instead the shell is rolled to true circular shape with the butt ends kissing but not overlapping, the seam is formed by external and internal butt straps, the sandwich being riveted together and importantly, true circularity is maintained. Fatigue fractures are not generally a problem with this mode of construction, but a phenomena known as caustic embrittlement raises its ugly head at times!

Boiler feed water treatment at one time contained plenty of caustic chemicals and whenever these seams suffer from persistent leakage, the chemical cocktail is drawn into the contracted spaces within the seam. Evaporation concentrates the chemicals which then aggressively attacks the steel. The attacked areas are generally the rivet holes, which suffer from radial cracking. The danger comes when the cracks link together between holes, the seam rips in an explosion.

The answer is careful inspection of the seams, giving particular attention to reported seam leakage. Before the advent of ultrasonic testing the follow up consisted of removing the rivets in the affected area, crack detection of the holes (magnetically or by dye penetrant), followed by appropriate repair.

This mode of construction has been completely superseded by the advent of the welded butt seam, the quality of which is upheld by vigorous metallurgical, ultrasonic or radiographic techniques. It is interesting that almost every technological advance throws up in due course some problem. Butt welded seams were no exception, after decades of boilers built in this manner without serious mishap, industry was shocked by several serious boiler explosions.

After much research and investigation it was found that the primary cause was lack of circularity. Steel plates are rolled to circular form using three roller bending machines. As the plates exit the machine the last few inches are propelled by only two rollers and the bending effect of the third roller is lost. Unless further work is done, flats will exist on either side of the proposed seam and true circularity is lost.

The presence of these flats called peaking, puts additional stress onto the welded seams, compounded by the fatiguing caused by varying pressure and temperatures within the boilers. As a result of these problems every boiler with this construction throughout the country was checked for the defect and remedial work done.

So much for cylindrical shells under internal pressure. With cylinders under external pressure, such as boiler furnaces, the effect is the opposite. Deformation or crushing is the problem, a perfect circle has high resistance to these defects but any departure from circularity is a source of weakness.

Furnaces are subject to intense heat internally and steam pressure externally. They are designed for this service, but if scale is allowed to form, then the steel has to endure much higher temperatures, ultimately causing softening and, under the influence of pressure, distortion and the assumption of an oval shape.

Once any degree of ovality occurs the furnace progressively distorts and eventually becomes dangerous. This type of defect is looked for at inspection times and the horizontal and vertical diameters of the furnaces are always measured. The difference between these measurements indicate the amount of distortion and where this is excessive, repairs are then required.

An extreme case occurs when a boiler runs short of water, then the upper furnace parts collapse completely, sometimes down to the line of the horizontal diameter and occasionally further. The great danger in this situation is that the furnace will tear and thereby initiate an explosion, the force of which has in some cases propelled the entire boiler out of the enclosing boilerhouse.

The repair of this type of mishap does full justice to the skill of boilermakers, if the distortion is slight, then by judicious use of hydraulic jacks the furnace can be returned to almost its original circular form. It is however weakened and distortion can be expected after a few years, major repair, similar to that required after a complete collapse is then required.

Major repair consists of cropping out the defective portion. The furnace is cut longitudinally at both sides and depending upon design, detachment from the riveted flanges or circumferential welds. A new portion of furnace is fitted, butt welded to the remaining original part and resecured at the ends by riveting or welding, thereafter checked by ultrasonic means and final hydraulic test. All this accurate work done in a very confined space - hot, fume laden and noisy.

The production of the piece of replacement furnace is quite a task in itself. A flat plate is rolled to the correct diameter and length and the seam butt welded. If the ends are to be secured by rivets, then each end is then heated and swaged over to form the flanged ends. A complete furnace ring is thus formed from which the required piece is cut out, ready for insertion.

***

18. Locomotive Type Boilers

THE locomotive boiler - like many engineering constructions - is a compromise between good design and practicality, refined by a long period of evolution.

No self-respecting engineer would design a pressure vessel having a rectangular firebox rather than a cylinder as a box is inherently weak and needs a wealth of secondary support to resist pressure.

Nor would he design a boiler that was so lacking in accessibility at the waterside - the very place that scale forms and defects propagate - but the boiler has to fit in a railway engine, be hand fired and have high output for its size, hence the compromise. No other boiler spends its life bouncing about on rail tracks, is flogged mercilessly, suffers wild variations in pressure and temperature. Even when sleeping it corrodes like fury, as rain and damp combine with acidic ash.

My first encounter with a locomotive was during training when I was sent out to Darwen during the town's holiday week. The address was a factory on the outskirts but on arrival at the gate house I learnt that the engine was elsewhere, about three quarters of a mile along the railway tracks at the engine shed. It just happened to be pouring with rain!

The engine was an 0-6-0 industrial tank engine and like almost everything else I inspected it was of at least 1910 vintage. It was placed over a deep inspection pit which of course had about 6" of water in it. Since none of the works' staff were about - it being the holidays - I had to float out a few baulks of timber into the abyss. The one redeeming feature was the boiler had been cleaned down to bare metal and a comprehensive inspection could be made with several pages of defects noted down.

Having carried out the inspection, I decided out of interest to have a look around the rest of the engine, the motion work like that on most railway engines was filthy with a mixture of ash, oil and coal dust hanging from threads of cotton waste like Christmas decorations from the various parts. Whilst I was having a look underneath, the vindictive beast took aim and hit me fair and square in the ear with what must have been its best globule, about the size of a golf ball, the filling ran down my neck and took refuge in my good shirt, about six back copies of The Daily Mirror found in the toilets removed the primary soiling or oiling, followed by the same number of cold water washes!

It was a good experience, I no longer sustained an emotional rapport with railway engines and I also decided that the lack of clients' support staff and of decent washing facilities would be a no go area in future.

In the late fifties when I started out in my district, the railway preservation movement was gathering momentum as was the desire to restore steam rollers and traction engines. My first encounter was a beautiful showman's' traction engine, bought by an enthusiast whose abundance in zeal was not matched by his technical abilities.

Years before I arrived on the scene, the enthusiast had been beavering away with bulk orders of welding rods restoring the years that the locusts had eaten, followed by artistry with a portable grinder, thereafter to be painted with lashings of red oxide. 'Restoration man' was at great pains to inform me that the boiler was now as good as new and that I need have no worries. Unfortunately I had one very big worry – how I was going to tell him that he had just about totally wrecked the complete boiler.

Breaking such devastating news is hard even for a diplomat like me - if I was starting afresh I am sure a course in bereavement counselling would help me cope with the odd restorer.

The problem with unauthorised repairs, particularly by welding, is that the original condition is unknown and cannot be determined, the plates could have been paper thin and no amount of welding could be relied upon to restore lost strength. The notion that weld material can be plastered on like repairs to a kitchen ceiling is completely wrong, it is the calculated strength or otherwise of the original remaining material that determines the manner of repair.

Sadly this engine was eventually scrapped, the owner realising it was now impossible to apply correct preservation techniques.

Restoring old railway engines, traction engines and steam rollers is a long, tortuous and expensive job, the secret is to seek advice at every stage, with it the job should run like clockwork. It will still be expensive, but without proper advice there will be many sleepless nights.

As a point of interest traction engines and steam rollers are classed as road locomotives since they share the same type of boiler and steam engine as a railway locomotive.

After about ten years further experience with locomotive type boilers I had acquired some very definite opinions on them and their zealous owners. The call for my services was not met with shrieks of delight. The archetypal first inspection would be in some remote windswept rickety shed, waterless, sinkless and tapless, an abundance of portable inspection lamps, only one of which worked but naturally somewhat lacking in electrical safety and insulation. The boiler propped up on bits of tree trunk, themselves half submerged in the requisite sea of mud, the enthusiast dancing round pointing out the highlights of his acquisition and its supremacy over similar traction engines of the same size. On informing him that to inspect the firebox I would need to get under his beloved but had no intention of attempting to swim the back crawl in mud, even though it was on such hallowed ground, our man produced a tarpaulin sheet to float out over the mud, fine apart from the few holes, presumably intended to speed up the inspection.

Having found the few dry parts on the tarpaulin and declining the offer of his dodgy lamp I start the firebox inspection with a candle, at least I can only get burnt which is preferable to electrocution. Despite the encircling gloom a thin shaft of sunlight is visible, now what a surprise, the firebox is corroded right through - must have been in the rain for the last fifty years.

The flickering candle reveals further corrosion, headless rivets and stays, cracks and distortion. It's too good to be true, this will keep our enthusiast occupied for the next ten years and off my back, I can see his feet from my cramped position under the firebox doing a nervous military two step as he awaits my verdict. I inform him with the utmost diplomacy that his firebox is kaput, finished, knackered beyond repair but ever the optimist he suggests he can weld every pit, crack and defect in the expectation that at the next inspection all will be well – in his dreams.

So, these were my thoughts when I received a call from a complete stranger, none other than Fred Dibnah who informed me he had a partly restored Aveling and Porter steam roller. My first thoughts were back to rickety sheds and seas of mud - must get on that bereavement counselling course.

My infamy must have preceded me, for much to my surprise every hitherto issue had been addressed. The boiler was in the dry, below was brown paper covered by a red stair carpet (perhaps he was expecting royalty!), the portable light was brand new and even boasted an earth leakage breaker.

The boiler could have come out of a show room and all the dismantled fittings were laid out on brown paper. His kitchen table was laid out with all the paperwork, drawings of the new firebox, steel maker's plate certificates, welder's certificates, radiographs, everything needed. If the boiler had a birth certificate it would have been there. As for the boiler, it could have just come out of Aveling and Porters' showroom, new firebox, barrel, smoke box tube plate and tubes, no sign of the restorers' favourite art form of pyramids of weld. Since I had attended at about 8.00am and the inspection took about two hours I allowed myself to be taken on a tour of Fred's empire, house, garden and machinery, managing to escape just before 5.00pm! It was a fascinating world, a great tribute to his innovation and persistence in overcoming major technical problems. Whatever the problem he was completely undaunted, be it forging complex components for his engine, pile driving for his house extension or even machining massive stone pillars. His joinery apprenticeship and also attendance at art school had instilled a great sense of accuracy and detail reflected in all his work, be it the profile of the humble rivet head or the brick and stonework in which he also excelled.

Our friendship extended over twenty five years, each inspection time was greatly extended to include a full explanation of his latest project or acquisition, such lengthy expositions belied the fact that he was a superb listener and observer, whatever he heard or saw would be stored and later repeated in some practical endeavour. He may not have had an engineering training but he most certainly had the mind of one, always searching out the how and why machines work, how they wear and go wrong, how they be repaired.

On my last day at work the firm arranged a retirement party, to my delight they had invited Fred to make the presentations, you probably know from experience how tight insurance firms are, the proverbial duck's backside will come to mind. One of my presents from the firm was unbelievably, a piece of exploded boiler tube which when weighed in was valued at 32 pence! I still have it but it's not on the mantelpiece!

19. Boilermakers – Industrial craftsmen par excellence.

THE job involves a close relationship with boiler makers, quite simply we inspect and they repair and for me it would be unthinkable not to give prominence to their craft.

Unfortunately the work of many industrial craftsmen is seldom seen by the general public as their work is confined within factory premises and doesn't see the light of day. Even the popular example of craftsmanship, the restored railway engine, barely exhibits such skills as most are hidden under lagging and cosmetic paintwork.

A further problem resides in the fact that whereas it is said that beauty is in the eye of the beholder it may be equally said that craftsmanship is in the knowledge of the beholder. Without some idea of the problems and techniques of a particular craft it is difficult to be fully appreciative of the immense skills used.

With these points in mind I must attempt to portray the craft of the boilermaker. First of all the title is very nearly synonymous with boiler repairer and my involvement was nearly always with the repair side. Many of the repair firms have a long history stretching back to the industrial revolution, the methods of repair have been refined over the years and are still evolving. There are few places where the boiler makers' craft is on show although industrial museums may have a boiler cut open to show its constructional details or sometimes a railway preservation society will have an unlagged boiler in its yard awaiting restoration. The observer may well ponder on for instance the skill and effort required to form the transition flanging that couples the rectangular form of the firebox outer plates to the cylindrical barrel, bearing in mind that it is generally of ½" thick plate and above. Or perhaps how such heavy and unyielding plate work can be made pressure tight.

A peep into the firebox itself will reveal exquisite flange work and row after row of perfectly formed rivet and stay heads, likewise with the expanded tube ends - perfect. But this is only a tiny fraction of their work, a trip into the industrial scene would reveal a myriad of pressure vessel types from immense refinery equipment down to the steam jacketed mixing vessel.

Repair work is to my mind the best indicator of their skill, cutting out cracked and defective plates, preparing replacement pieces, very often involving intricate flanging, marrying them up to existing plates - work that calls for immense skill – then securing by either welding or riveting. Welding or riveting at a bench is one thing, transfer it to the inside of a vessel cramped up, lying under the work site, welding rod in hand, showered with red hot sparks inches from the face and engulfed in flux fumes is quite another experience altogether. After each of many weld runs, grinding out to form a sound base for the next run of welding is another hazardous operation, this isn't just welding to close a gap, it is welding to impart at least the same mechanical strength as the plates being joined. All such strength joints are subject to ultrasonic or radiographic test, despite the appalling working environment the job is invariably perfect.

Artists and craftsmen must be allowed a little eccentricity, my first and many subsequent encounters with boiler repairers required me to explain this to several irate factory managers. First indication of repairers would be the sight of an old van, for some reason invariably a Bedford or Commer heavily listing to either port or starboard and down by the stern under an overload of steel plates, gas bottles and other tools of the trade, the abused vehicle laboriously towing a four wheeled diesel compressor or welding set.

The very first activity of the squad of four or five men would be the setting up of base camp, generally out of the way in a corner of the boiler house, the centre piece being a coke or gas fired portable furnace around which would be the tea cans bubbling away.

The squad sitting around the brazier on boxes and the odd sequestered chair, would be in earnest discussion on the merits of horses for the 3.15, and in due course the labourer would be dispatched to the local bookies.

All these activities caused huge frustration and irritation to some factory managers who turned to me for an explanation, I took the line that before starting work it was imperative that a full discussion on the method of repair was necessary. Trust them, they're craftsmen. After the job is started the odd visit to base camp for recuperation is fully justified after sweating it out, bent double, riveting or welding in some contracted space.

In my latter years at work, now as a manager, the equivalent of base camp was endless meetings, seminars and conferences, held in salubrious hotels with all the trimmings.

Unlike the brazier forums not much was actually achieved. I cannot recall horses with all the promise of winning ever being on the agenda, apparently our equivalent was to be noticed as we eyed the next rung of the ladder of success!

In a previous chapter mention was made of furnace defects where distortion occurs and requires repair, this is another example of the boiler repairer's craft, using hydraulic jacks in the confined space, seldom more than three feet in diameter they manipulate the furnace back to circular form. Many boilermakers are in need of serious repair themselves, a finger count seldom reaches ten and the one eyed artisan is not a rarity as it is a most dangerous occupation.

***

20. Off the Peg – On the Blood

WHEN visiting preserved steam plant, very often its size, proportions and general appearance captivate the eye and we forget that there is plenty of interest in the details.

For instance, even the pressure gauge has a story to tell. It is usually fairly prominent, if on a boiler it will be high up on the front, generally it will be highly polished and our gaze will be directed to this focal point.

A close look will reveal that the graduated dial is marked at a certain pressure in a distinctive colour. This is a legal requirement and the mark will be at the graduation corresponding to what is known as the maximum permissible working pressure.

This pressure is determined by calculation based on the boiler's present condition and appears in the current thorough examination report.

The most common colour is red, although any colour of the rainbow will do, provided it is distinctive relative to the colour of the numbers and graduations on the dial.

The gauge has to be suitable for its intended duty, it is good practice for the total range of indicated pressure to be considerably greater than the maximum pressure reached by the boiler hence the red mark appears around the mid point on the dial. This ensures that the gauge works reliably well within its total operating range.

Suitability also embraces such matters as size and being within easy sight of the operator, it has to be kept clean and if in a dark place then illuminated.

Many of the gauges, particularly on old plant, are well worth a second look. At least one maker arranged for the operating mechanism to be on view and this was achieved by having an annular rather than a full face dial.

Most gauges utilise a bourdon tube which under the influence of the pressure to be measured distort and provide the movement to actuate the pointer or arrow. This tube operates in the same way as a child's party horn. Here a flattened paper tube wound in a spiral, scrolls out when blown into, the bourdon tube however is of metal construction.

The small movement of the tube when subject to pressure is magnified by a system of levers and cogs which rotate the pointer, most of these parts are of lacquered brass or stainless steel and do full justice to the skills of the instrument maker.

The pointer itself is often ornate, depicting an arrow, its centrally placed boss is arranged to be a tight fit onto the actuating spindle, hence its security relies on friction. At the zero graduation a small peg protrudes from the dial to prevent further anticlockwise rotation of the pointer.

Occasionally when a boiler is observed cold and out of use the pressure gauge indicates a slight pressure present, sometimes as high as 10psi which in reality cannot be correct.

This anomaly can be explained by the sequence of events when a boiler or for that matter any other steam vessel is taken out of service and allowed to cool down. The cooling process lowers the steam pressure which eventually drops to zero or atmospheric pressure. The remaining steam then condenses back to water, which lowers the pressure further and a partial vacuum is formed within the vessel.

This remains for a short time but eventually air leaks into the vessel and the internal pressure returns to zero or atmospheric.

These events have an effect on the gauge, as the pressure drops to zero the pointer responds and eventually reaches and rests on the peg, as the pressure then drops further the pointer spindle responds but cannot move the pointer due to the peg. If insufficient friction exists between the pointer and the spindle then relative movement takes place and the accuracy of the instrument is compromised.

When the internal pressure finally returns to atmospheric pressure the pointer, now slightly deranged, incorrectly records a pressure slightly above zero. Generally this does not matter too much as the error is on the safe side.

Although recording steam pressure, such gauges are too delicate to be in contact with steam and its associated high temperature. Careful observation will reveal that the connecting pipework will incorporate a U-tube or syphon device and this feature will trap a quantity of water which will allow pressure but not steam to reach the gauge.

The chapter title is from the world of marine engineering. Raising steam from a cold start on boilers takes time, the first indication that progress is being made is when the gauge pointer leaves the peg, hence the expression off the peg.

On the blood – is the expression to denote that the boiler pressure is at its maximum, just below safety valve setting and the blood relates to the red mark on the dial.

With well-maintained safety valves no steam leakage is observable at the waste pipes until the pressure is on the blood. At that stage tiny wisps of steam may be seen, this is termed feathering and is the precursor to a full and noisy discharge. This wasteful event is acutely embarrassing to marine engineers and is avoided like the plague.

***

21. Vacuum on the Rampage

SURPRISINGLY, many pressure vessels are damaged by vacuum with the classic casualty being the steam heated calorifier. Designed to provide supplies of hot water, it usually consists of a cylindrical shell of thin copper with steam heated coils in the lower part.

The vessel is supplied at the lower end with cold water from an overhead tank and the effects of water expansion and the need to displace air on initial filling being met by a vent or expansion pipe situated at the top and extended to terminate over the feed tank.

Water temperature is raised by the internal steam heated coils and kept at a preset temperature by a thermostatic control valve in the steam supply.

This type of plant usually gives years of trouble free running, but there might come a day when the thermostat fails in the open position, the stored water then continues to heat up till it boils. This is not too serious as the generated steam escapes up the vent pipe.

But the situation becomes a catastrophe when a heavy demand for hot water is made. The exiting hot water is replaced by cold water from the overhead tank causing the instantaneous collapse of the steam bubbles within the boiling water. This causes a partial vacuum which can only be prevented by the rapid influx of air through the vent pipe acting in reverse.

If the vent piping is tortuous and of small diameter the inertia of the water within the pipe cannot break the vacuum in time to prevent the calorifier collapsing inwardly under external atmospheric pressure.

In these circumstances the calorifier can be totally wrecked, a squashed Coke can comes to mind!

An interesting case occurred at a local colliery bath house where the horizontal calorifier was the largest I have ever seen, being about 20' long by about 12" in diameter. It needed to be this size to meet the washing needs of several hundred miners at the end of shifts. It had collapsed and on first appearance was a total write off and would need replacement.

To have a new one built and installed would take weeks, meanwhile the miners lacked hot water washing facilities. An engineer at my office came up with a unique suggestion of attempting to restore the original cylindrical shape by the reverse process, that is, by pressurising it with water in the same way that a crumpled paper bag might be inflated. Even if unsuccessful nothing would be lost.

Accordingly, all the connections were blanked off and since the town's water pressure was about 70 p.s.i. - far in excess of the calorifier's normal working pressure - water was slowly and carefully admitted to the wreck.

Like a child's balloon the calorifier slowly regained its original shape, not perfect, but sufficient for it to operate for a further ten years until the colliery closed.

This type of repair is only feasible if the collapsed vessel is free from kinks and from excessive distortion at pipe connections and manholes.

Copper shelled textile drying cylinders can also be dramatically reshaped by vacuum on the rampage. These cylinders are usually about 10'long and about 15" in diameter arranged in groups of 10 to 20, all are gear driven.

Wet cloth transports over the rotating cylinders and in doing so is dried by the steam within the cylinders, the incoming steam enters and condensate exits each cylinder by means of hollow trunnions, the working pressure seldom exceeds 15 p.s.i. If the steam supply is interrupted whilst the cylinders are rotating and carrying wet cloth the contained steam rapidly condenses creating vacuum conditions which on occasions causes collapse and destruction of the cylinders. The cylinder design incorporates internal strengthening hoops to resist vacuum conditions and also vacuum breaking valves, the hoops can get displaced and the valves sometimes seize hence the cause of collapse.

These cylinders are not repairable in the same way as calorifiers as accurate circularity is required, the only possible repair is replacement of the shell.

Occasionally vessels and storage tanks collapse due to either pumping out or draining by gravity with a defective vent system, so inducing a partial vacuum.

Incidentally, it appears that the likelihood of mishap to pressure plant is inversely proportional to the normal working pressure. The lower the normal working pressure the more likely a mishap. There are several reasons for this including the fact that low pressure plant is constructed of thinner materials. Any defect such as a crack is of much more significance than the same defect in thicker materials. Low pressure plant is also generally supplied with steam or other fluid from a higher pressure source. This involves pressure reducing devices which sometimes fail causing over pressurisation and explosion.

The basic insurance cover for any form of pressure vessel is explosion and collapse, it is the latter upon which most claims are lodged.

***

22. Cracks and Fractures

THE harbinger of many engineering mishaps and disasters is a microscopic crack. The early detection of such cracks is therefore of paramount importance so that remedial action can be taken or so the defect can be closely monitored until repair is required.

Broadly speaking, cracks are the result of either simple overstress or a more complex phenomena known as fatigue. Engineering structures are designed with permissible material stress levels well within the limits of safety and it is exceedingly rare to witness failure due to a design fault causing excessive stress.

However, in the life of a machine or pressure vessel, when deterioration takes place stress levels rise which ultimately can cause cracks and failure.

Examples include bolted structures in which one or several bolts slacken throwing excessive stress on those remaining, Failure of one or more supports in a multiple support system or perhaps corrosion causes a general weakening of the structure.

The classic overstress crack leading to catastrophic fracture sometimes occurs in the bottom end bearing of large engines and air compressors. Here the two halves of the bearing are united by just two bolts, the slackening on one throws all the stress on the second bolt which promptly fractures under the double load.

When the wrecked engine is examined the bolt that initially slackened will be found to have bent into a U-shape allowing the bearing halves to open out. Careful examination of this bolt will reveal the tell-tale signs of slackness - indentation marks under the head and under the securing nut.

Overstress in pressure vessels is fairly rare, bolted lids and covers such as used on autoclaves, kiers and dying machines are sometimes abused by operators not schooled in safety. Finding a bolt or nut too tight to screw down, it is left unused, greatly increasing the stress on the remaining bolts thereby inviting fracture and dicing with the possibility of explosion.

The most common crack is however occasioned by fatigue. Surprisingly, materials can get tired, not with constant static stress but with varying stress. The greater the frequency of changes in stress the more likely failure from this cause. The bored office worker idly bending back and forth the long suffering metal paper clip, which eventually snaps, is a reasonable demonstration of fatigue.

The magnitude of the dynamic load inducing a fatigue crack is much less than a sustained static load, the office worker could never be able to apply sufficient static load to cause fracture.

Every crack of either type originates at a specific point of high stress, that is, relative to adjacent moderately stressed areas. Points of high stress are caused by abrupt changes in surface profile. It could be a microscopic scratch on an otherwise highly polished surface or a machined sharp-cornered shoulder in a piece of shafting. Cracks will preferentially initiate from these areas.

With large complex items of plant it would be virtually impossible to minutely examine every square centimetre of the structure, fortunately there is over a century of accumulated experience to draw upon, we are well acquainted with the causes and likely sites of such defects.

The inspecting engineer faced with a large complex vessel such as found at chemical works or refineries will first obtain an overall view of the vessel's function, the pressures, temperatures, fluids used and so on. Imagination plays a part, the ability to visualise the effects of pressure and heat, the frequency of these activities and where the effects will be concentrated, all help to narrow down potential problem areas such as local geometric shape and compromised surface finish.

Shafting, including engine crankshafts, are classic

subjects for fatigue, in fact anything that rotates or reciprocates is a potential problem. At one time newly completed vessels were stamped with the manufacturer's details until it was noticed that sometimes cracks emanated at one of the stamped letters leading to full blown fracture. Such stamping is now restricted to areas absent of dynamic stress and of low static stress, attached maker's plates being preferred.

The ancient beloved Lancashire boiler (not on my Christmas present list!) was a great place for fatigue crack-hunting. The flanged front end of the first furnace tubes were to all intents and purposes married to the goddess of fatigue. Taken into any such boiler after a night on the tiles and blindfolded, the cracks could easily be found and felt with a bandaged sore thumb!

The furnace tubes in such a boiler are at least 25' long, expand and contract longitudinally in response to the varying temperatures and pressures. This movement is concentrated at the front furnace flanging – just waiting to be fatigued.

Flanging in general is a good antidote to fatigue as the stresses are spread over the curved flanged area. But the constructive process is by rolling and tooling, inevitably this leaves the surface finish imperfect and the goddess alights on some imperfection.

The best is yet to come, the water within this type of boiler has been or is usually hostile and aggressive, acidic to the purist. The microscopic crack exposes fresh metal crystals to the acid which starts a corrosion process. The crack mouth rapidly widens and within months the crack can be 1/16" wide at the surface. Generally speaking however, this form of fatigue crack propagates at a snail's pace, the corrosion a little faster than the crack advance and often it takes ten years to become of significance.

When the crack lengthens to about 21", a horrific sight to those not in the know, we start to contemplate repairs. Occasionally multiple parallel cracks appear like a ploughed field.

In contrast, a minute crack in an all welded boiler or pressure vessel, calls for rapid if not immediate repair. Welded structures generally have less flexibility than riveted structures and cracks advance at alarming speed.

This chapter has been about finding cracks by sight and experience, but the findings are limited to the crack length and width at the surface. This amount of information is sufficient in most instances but in some situations we need to know more, the depth, direction and rate of propagation – this requires another chapter.

***

23. Rat up a Drainpipe

MILDLY claustrophobic? Then best skip this chapter, for this is all about entry and exit from very confined spaces. The thorough examination of most boilers and pressure vessels requires at the very least visual inspection of the external and internal surfaces and both can give access and egress problems. This area of activity has for many years been covered by progressive legislation.

The provision of an opening in a pressure vessel to permit access weakens its structure and therefore has to be as small as possible, the weakening is compensated by the provision of any extra layer of plate or substantial frame around the hole. At the next visit to an industrial museum keep a lookout for these compensation devices!

The actual holes are generally elliptical or circular, the former generally 16" x 12" and the latter 18" diameter. Both types are fairly easy to traverse provided that that are no obstructions on either side of the hole. Occasionally vessels are fitted with 15"x 11" elliptical manholes and personally I would like to meet the designer in a back alley on a dark night!

The elliptical shape has two advantages, firstly, it closely resembles the cross section of the human body and therefore permits a design with the least possible size of hole. Secondly, the elliptical shape allows the manhole cover to be passed through the frame opening into the vessel, this is advantageous as the pressure within is utilised to force the cover against the frame, thereby assisting the gasket to achieve a tight seal.

The cover is spigoted to ensure correct alignment with the frame and crucially to prevent pressure ejecting the inserted sealing gasket. It is also provided with two substantial studs. With the cover in the correct position, strong backs are passed over the studs and nuts are used to draw the assembly tightly together.

Poorly maintained elliptical manholes and their covers have been the cause of many fatalities, particularly with steam boilers. Apart from the usual wear and tear problems associated with nuts and studs, it is vital that the clearance between frame and spigoted lid is kept to a minimum.

If the gap or clearance exceeds the gasket thickness then the possibility exists for the gasket to be explosively ejected, followed by the rapid discharge of the boiler contents.

There are many considerations concerning entry to confined spaces, paramount is personal safety, Is the space properly ventilated, free from dangerous chemicals, are mechanical stirring devices electrically isolated, are all valves controlling the admission of fluids to the vessel securely locked shut or better still blanked off?

Also important is the the presence of person on the outside to give assistance in case of difficulty. Steel boilers and vessels are subject to rusting when out of service, this depletes the oxygen content within the vessel and therefore it is important that thorough ventilation is achieved before attempting entry.

The writer was once about to enter a steam jacketed mixing vessel, fully assured by a works' staff that an internal stirring device was electrically isolated. However, when asked to prove the supposedly safe situation by attempting to start the driving motor, to everyone's surprise the stirring blades whirled round. Investigation found that the electrician had isolated the wrong motor – a good lesson, always test the veracity of steps taken to ensure safety.

As I was about to examine a power station boiler I made the usual and important check that all valves isolating the boiler from other equipment were securely locked in the closed position or blanked off. At this particular station the method of satisfying the law's requirements involved shutting the valves and securing the hand wheels with a locked chain to prevent rotation. On this occasion however, several of the locked chains were too slack to prevent rotation. It was most unlikely that anyone seeing the chains would attempt to open the valves, but it did indicate a sloppy unsatisfactory approach to safety – what other short cuts were being taken?

An amusing situation once occurred at a pharmaceutical factory, the item to be inspected was a large glass lined reaction vessel used in the production of medicines and drugs. Before anyone is allowed to enter the vessel a chemist takes air samples from within and after testing the air quality and finding it satisfactory, produces a safe to enter certificate.

After perusing this, I was told to wait as a member of staff called 'Billy' would accompany me into the vessel. In due course a man arrived holding a tiny cage, ' Billy' the mouse was the resident!

I was instructed to keep a close eye on the mouse during the inspection, if it rolled onto its back with feet in the air, it was high time for me to make a swift exit!

Despite sophisticated air sampling techniques it is possible in a complex vessel to have sludges remaining in contracted places giving off noxious fumes not picked up by sampling. The use of a mouse, like the canary in mines, is a very good safety precaution.

Despite being fully acquainted with the dangers of entry to confined spaces I broke the rules on one occasion and paid the price. The boiler in question was a very small horizontal tubular type, never intended to be actually entered, my zeal far exceeded experience. After several attempts I managed to squeeze in, once inside I soon realised I was wedged like a sardine in a can, could hardly move yet alone carry out any meaningful inspection.

In a less tight situation the technique for exiting would be to place hands on the manhole mouth and heave oneself out, in this case it was impossible to exert sufficient force. My second mistake was not to ensure a member of the work's staff was available, it was in fact lunchtime. Fortunately, I am not a signed up member of any claustrophobic club. So I settled for a rest on a bed of tubes and waited an hour for some of the works' staff to arrive, they with much heaving and pulling managed to get my head through the opening. The rest of me was pulled through like a cork from a bottle, a little sore and definitely much wiser!

***

24. Fittings and Supplies

THE safety of the boiler or vessel having been inspected and found to be in perfect condition still hangs on a thread, or more precisely on a fittings and attachments.

The safety or relief valve being the best known, it has to prevent the safe working pressure being exceeded. For this it has to be set at or below this pressure, has to be in proper working order and has to be of sufficient discharge capacity. Infrequent use very often spells problems, failure to operate or occasionally failure to shut after operating.

The writer requested a boiler safety valve test at a large dye works, it opened at the required pressure but failed to shut and remained open for over an hour, discharging so much steam that the entire works had to shut down. It was over another half day before the boiler cooled sufficiently before the problem could be investigated and remedied - no doubt I was not the flavour of the month with the production manager!

The maintenance of a safe water level is paramount in boilers, failure results in serious damage at the least and in the worst case explosion, there are statutory requirements in place to prevent these and other mishaps.

At one time the boiler fireman exercised manual control over the supply of water and in the event of mistake an internal float device signalled low water by a discharge of steam – a reasonable solution, considering the technology of the time.

Today, it is electronics in conjunction with float or probe devices, very often these are contained within an external chamber connected by pipework to the boiler itself, a high standard of maintenance and frequent testing is the order of the day.

Most steam and compressed air operated equipment is supplied from a much higher source of pressure, this brings its own dangers, failure to limit the supply pressure by properly maintained devices has been the cause of many explosions.

Reducing valves drop the pressure to the required limit. It is a statutory requirement that this is followed by a suitable safety valve, so set that the safe working pressure of the equipment cannot be exceeded in the event of the reducing valve malfunctioning.

Most factory managements are highly responsible in these matters, but one unfortunately suffered an explosion due to of all things, paperwork errors.

I was called out to an explosion of a stentering machine, the caller used the apparently standard phraseology to acquaint me with a mishap: "The stentering machine you inspected on such and such a date has exploded!'

Presumably the linkage of my previous inspection with the mishap is taken for granted - in their dreams!

Three weeks previously, during the town's annual holidays I had indeed inspected the machine, in essence a rotating cylinder about 8' diameter and some 4' wide, the sides or ends being of steel whilst the cylindrical shell was made of thin copper, working pressure 30 lbs/sq. in., date of make around 1910.

If you ever visit a textile museum these vessels together with any similar steam heated drying machines are well worth a second look. The attachment of the copper shell to the steel sides underlines the coppersmith's skills. By judicious flanging and manipulation the copper is married to the steel and clamped with a bolted ring. The longitudinal lap seam in the shell is secured by flat headed copper rivets, solder is floated over the seam to enable a smooth transit of the cloth. Rotating thin shelled cylinders in the textile trade seldom explode but frequently suffer implosion or collapse, caused by vacuum conditions when the steam is shut off and wet cloth is still on the cylinders. To prevent this type of mishap the cylinders are provided with vacuum breaking valves at each side, however, the close proximity of cloth and fibres occasionally seizes these valves.

The tenter in question was at the inspection found to be in good condition but the steam supply arrangements required urgent attention. The steam supply was from boilers working at 180 lbs/sq. in, reduced to 30 lbs/sq. in.

The reducing valve was definitely of pensionable age and the safety valve was seized solid, accordingly I saw the works' engineer and left him with a note to the effect that the machine should not be used until completion of repairs. Unfortunately, after I left the premises the engineer suffered a heart attack, again no linkage to me!

He and the note were carted off to hospital, however it would take several days before the clients would receive an official Factories Act report – no laptops in those days.

The usual route for such reports on arrival at the works would be for the manager's secretary to forward it to the engineer who after dealing with it would return it to the office for insertion in what is known as the general register, this being a legal requirement enabling perusal by visiting factory inspectors.

Being the town's wakes week the manager's secretary was on holiday and a temporary secretary was standing in, she looked at the form which always carries the heading – *To be kept in the General Register* – and understandably did just that, thereby depriving the engineering staff of the vital safety information.

After the holiday shutdown the works started production without repairs to this machine. Shortly afterwards the reducing valve failed to control the steam pressure and the seized safety valve could not relieve the excess pressure. The stenter, now grossly over pressurised, exploded in spectacular form. Most of the copper shell was ripped off - an opened sardine can comes to mind - and all the windows were blown out of the stenter building. The shocked and startled operator was physically unhurt but was not seen around for a few weeks!

Another works suffered an even more spectacular series of explosions contingent on reducing and safety valves failing to operate. The damage in this case was to the entire length, some hundreds of feet of large diameter low pressure steam piping. The works was shut down and out of production for two weeks whilst all the piping was replaced.

In this case the source of steam was from boilers operating at 350 lb/sq. in. reduced to 30 lb/sq. in. protected by two safety valves situated on the top floor of a tower. The valves were of stainless steel and after the mishap were examined and found to be completely seized due to chemical deposits on the working parts - a most unusual cause. Further investigation revealed that the tower acted like a chimney and had a considerable updraught. At its base caustic chemicals were being used, the fumes from which were drawn up into the valves and caused the problem.

Regular testing of the valves would have shown up the problem and prevented a mishap, understandably works' staff are often reluctant to carry out periodic tests as such testing causes interruption to production, albeit minor compared to the time spent on repairs if a mishap occurs.

***

25. The Flying Saucer

INDUSTRIAL dyeing involves the use of large vertical pressure vessels, often several feet in diameter and height. The top is a hinged lid in the shape of an inverted saucer, secured by six or more swing bolts.

Another design has a bayonet type fastening. Here the lid periphery and the vessel top have matching castellations and a slight rotation of the lid secures and locks the assembly, in the same way as a light bulb fits into its holder.

The vessel is connected to an external circulating pump and either in the lower part of the vessel or in the associated pipework, steam heating coils are fitted.

After material to be dyed has been loaded into the vessel and the lid secured, the system is filled with water based dye, circulation is commenced and by means of the steam heating coils the water temperature is slowly raised.

During this initial heating process the contained water expands and a small tap or valve on the vessel lid is left open to relieve this surplus water.

When boiling point is reached the tap is shut, the pressure and associated temperature rise to the predetermined value of about 70 P.S.I. which equates to about 310 degrees Fahrenheit. On completion of the dyeing process the vessel is drained, the lid opened and the material removed.

The vessel and pipework are protected by at least one fairly large safety valve set at just above the working pressure, in the normal course of events these valves are not required to operate and therefore should be tested at least weekly.

A devastating explosion occurred at one dye house when the whole lid assembly parted company from its vessel, taking part of the dye house roof with it and landing 300' away, but fortunately nobody was injured.

Subsequent investigation of the substantial stainless steel lid found it to be badly distorted around the bolt landing brackets, the bolts themselves stretched and bent. Calculation showed that the pressure required to inflict this damage was many times the normal working pressure. The only way this excessive pressure could be generated was by the thermal expansion of the water during the heating up stage - the top drain tap must have been shut much too soon. This mistake should not have resulted in an explosion, the safety valve should have relieved the excess pressure.

This valve was tested and found not to operate at ten times its designed pressure, with some difficulty it was dismantled and found to be seized solid. At some time during an overhaul one of its internal parts had been badly distorted, probably by use of an incorrect tool, it certainly had never been tested after the overhaul.

Generally speaking clients are reluctant to initiate the regular testing of safety devices, the maxim of 'if it isn't broke why fix it?' is all too common. Safety devices can only be proved and kept in working order by frequent testing.

The insistence of testing at specified intervals, signed for by a staff member, works wonders in these cases.

***

26. Candle to Calculus

THE tools of the trade start with the candle - and probably end with calculus! While the usual engineering measuring devices are used with some adaptations, this chapter looks at some of the more specialist equipment in use.

The effects of corrosion and wastage are seldom sufficiently uniform to enable an accurate assessment of the remaining strength of materials to be calculated, the area as well as the depth of defects has to be taken into account.

With this information, a measure of experience and with a bias towards safety we can come to the correct conclusion.

With very old plant the quality of the steel used and the working conditions endured very often gives rise to such seemingly haphazard corrosion. This often occurs in the contracted parts of boilers and vessels. It is here that on occasions a candle is better than a torch.

The flickering light source suitably positioned on the end of a length of bent coat-hanger wire returns a view of moving shadows from which the profile and depth of a defect can be fairly accurately assessed, whereas torchlight merely returns a two dimensional, deadpan scene.

The early detection of fine cracks has always been of interest to engineers and with oily machinery advantage has been taken of the propensity of oil to penetrate cracks by capillary action. The suspect surface is wiped clean and dusted with chalk, oil seeps out from any crack present into the chalk and gives a visible indication of its presence.

This technique has been sophisticated courtesy of the pressurised spray can. A detergent spray is first applied to thoroughly clean the suspect area, a dye, usually red and carried in a thin solvent is applied next and this penetrates any crack present. After a suitable time-lapse the area is again cleaned to remove all traces of dye, thereafter a chalk substance is sprayed on and dye emanating from any cracks becomes visible.

A variation of this method is to magnetise the suspect area and at the same time spray on a fluid containing microscopic iron filings. The magnetic field will cause the ferrous particles to migrate and bridge any cracks present.

All these methods are of limited use, as no indication of crack depth or direction is given. If this information is required then radiography or ultrasonics will be needed.

Radiography provides a permanent record not only of crack dimensions but almost every other type of defect, particularly those found in welded seams. However it is relatively expensive and on account of the shielding required, not always easy to apply in the field.

Ultrasonic testing is now in most cases a better option, eminently portable and relatively cheap. The slight downside being that the information received is in graphical form, requiring conversion by the specialist operators to dimensioned drawings.

A simpler hand held ultrasonic device for merely checking plate thickness is widely used, pressing a probe onto the plate gives a digital readout accurate to much less than a millimetre.

Finally the most useful tool of all, the calculus, without which our understanding of the behaviour of flat plates for instance under pressure would be very limited. With it, we can precisely calculate and anticipate deflection, essential information in the quest for safety.

***

27. The Temple of Meditation

SEWAGE normally flows down by gravity from source to treatment works, but occasionally the local topography precludes this method of transit, such as where the source is lower than the main sewer.

Generally, this occurs in remote rural areas where houses and other buildings are sometimes many feet below the main sewer.

A piece of pressure equipment known as a sewage ejector chamber solves the problem. The sewage flows by gravity from these local areas to the ejector where it accumulates in the vessel. An internal float signals when the vessel is full and at this point compressed air is admitted to the chamber. This forces the sewage out into piping connected to the main sewer.

On completion of the discharge the chamber is vented to atmosphere and is ready to refill, the cycle is entirely automatic.

Like all mechanical equipment associated with civil engineering works it is robust, substantial and very long lived. Because it is a pressure vessel it is however subject to periodical inspection, but I have never found a serious problem with them.

Most such devices are installed within a brick or concrete structure at least 10' below ground level to ensure good flow velocities. The building above the underground installation houses the air compressor system, also subject to inspection.

Ejector chambers are in fact a substantial iron casting, the preferred material to resist corrosion, several inches thick and cylindrical, the lower portion being of an inverted conical form. A heavy lid is bolted to the top, removal of which permits entry for inspection purposes and maintenance of the float mechanism.

I do not envy those tasked with preparing this type of plant for inspection, any remaining liquid sewage is first discharged by the compressed air and after venting, the heavy lid is unbolted and lifted clear by chain blocks. It hangs above the chamber like the sword of Damocles, festooned in unmentionable stalactites although these are removed as much as possible.

The chamber is then entered and thoroughly cleaned down to bare metal, not a job for the faint hearted or squeamish, I liked to think that at least a drop of disinfectant was used at this point!

It is a fact that the time taken to prepare plant for inspection far exceeds the time taken for the inspection, embarrassingly so. In this case the preparation takes say, three men about 4 hours, whereas the inspection, due to the special circumstances, takes about 4 minutes flat!

To prevent embarrassment, Parkinson's Law is adapted, inspection time is expanded to meet clients' expectations. The entire scene is awkward, the cleaning squad squatting on the ground outside, filthy and smelling to high heaven are awaiting my arrival.

To lengthen the visit, I initiate and prolong conversation about everything from government to weather, when conversation runs out I reluctantly don a pristine white paper boiler suit and descend the ladder to the chamber.

Once inside, acclimatised to the terrible odour and sitting fairly comfortably, suitably insulated from the freezing cold cast iron by a thick wad of rag - for fear of piles - the inspection can commence. An area of apparent corrosion at my feet turns out be something quiet different when struck with a hammer, not altogether to my surprise.

After a few minutes the inspection is over and Parkinson's requirements are to be followed, at least a further 15 minutes must be spent in the chamber, on one such occasion I turned to meditation.

This is a great place to consider status, as one sits in

the place normally occupied by about half a ton of raw sewage, thirty plus years of endeavour and I had reached the bottom so to speak.

With these depressing thoughts in mind, I received a huge uplift, when considering the career of a fellow school pupil, his right hand permanently aloft, ever ready with the perfect answers, much to the chagrin of his average classmates.

On leaving school our paths diverged. he to Cambridge and I to the North Atlantic, however we both eventually arrived at similar venues, I in my chambers and he in his chambers – as a high court judge!

With these inspiring thoughts in mind, I left my chamber, mounted the Dagenham donkey mark two and sped home for a much needed bath.

***

28. Payment in kind

IN the 1970s many textile mills were closed down and then let or sold for use by other industries. One such mill I remember was bought by a paper converting firm who produced such items as kitchen rolls, toilet rolls and various kinds of wrapping paper.

The mill originally had four Lancashire type boilers providing steam to a large engine. The paper firm scrapped three of the boilers and the engine as the remaining boiler easily met their requirements for heating and process steam. Due to production requirements the boiler could only be shut down for inspection requirements at the weekend which in those days attracted a fee much welcomed by myself.

On reminding the works' engineer that the boiler was due for examination he arranged for the boiler to be ready in a few weeks' time on a Sunday, when the fee was mentioned he promised to get back in touch. A few weeks later he phoned to say he had spent the year's budget and there was no money for fees or much else come to that. I reminded him that he had a choice, shut down in the week and not have the additional fee or have Sunday and pay the fee – a rock or a hard place.

Two months later I had to remind him that the boiler had become legally overdue for examination and apart from possible legal consequences in the event of a mishap we, as insurers, may not look too favourably upon any claim that would be made. He came up with a solution, would I inspect on the next Sunday and be happy to be paid in paper – toilet rolls!

Fine by me, but how many? His reply, how about a boot full? I had just changed cars from a Mini to a Ford Corsair – enormous boot. Anyhow he kept his word and whilst I inspected the boiler he filled the boot to the lid, additionally filling the back and front passenger seats with rolls up to the roof, bit embarrassing for the drive home! The rest of the day was spent filling the loft with toilet rolls, it must have been about five years before I had to buy further supplies.

***

29. Convents and semantics

WHEN mishaps occur, repairs are required or liability has to be discussed, it is sometimes necessary to explain the situation to people not familiar with the plant involved.

One such case involved a large heating boiler in a convent which was leaking. This type of boiler has been partially described in a previous chapter and a little further explanation is now required.

The various sections of the boiler are joined to adjacent sections by short pieces of pipe known as barrel nipples, the barrel shape engages with tapered holes in the sections and on being pulled hard into each other by tie rods a water tight seal is obtained.

During the summer months' with the plant out of service, condensation and rusting occurs, mainly due to the hygroscopic nature of any remaining combustion deposits. Amongst other places the rusting takes place in the confined spaces between the sections and, like the action of ice, it imposes considerable force and when the force exceeds the bond of the nipples, leakage occurs.

On arrival at the convent I was ushered down to the boiler, a quick inspection revealed that at the 7th and 8th sections the lower left nipple was leaking sufficiently to impair working of the boiler, the lower right and the top nipples were in satisfactory condition. The clients had an extended policy which covered nipple leakage and in this case we were clearly liable.

When talking to most people and especially engineers where there appears to be some slight emotional attachment to the plant within their care it is customary to start the conversation with such words as "Your second tube, row 4......" " or "You've got a problem with......".

My next move was to have an audience with the Mother Superior to inform her that her problem was leakage from the left nipple, none from the right or top nipples and that we were liable for repairs. The mention of nipple leakage is bad enough, but the thought of a third nipple would be one too far!

During the slow walk up to the office I had ample time to rue my lacks in education as I wrestled with alternative acceptable wording. It came to me at the last minute, nipple leakage would be morphed to connector leakage. The conversation proceeded well till the Mother, fingering the insurance policy, stopped me in full flow, to ask where the risk of connector leakage occurred in the policy, with some trepidation I explained the 'Official' wording!

At the close of my visit she with a twinkling eye congratulated me on my handling of the affair, but thankfully was not too specific.

To my mind the wording sounds like the work of some wretched engineer with no thought for the embarrassment he might cause his brother grease monkeys!

***

30. Hands on! Hands off!

MOST types of plant require two distinct kinds of inspection: the first when it is at rest and the second when working. With pressure vessels subject to legislation these are known as the thorough or cold examination and the working or hot examination respectively.

The thorough examination is very much hands on, the old Lancashire boiler for instance involved a crawl on hands and knees of about 200', very often in temperatures of over 100°F.

If adjacent boilers were under steam then the examination of the side flues was really unbearably hot and gloves and knee pads were essential - however for the enthusiastic it was a good place to find problems with the circumferential shell seams.

Parliamentarians responsible for legislation have described this as the cold examination, maybe they were on another planet, but unlike them I happen to be earth bound and I've had the blisters to prove it!

The working or hot examination follows, the primary purpose of which is to prove the efficacy of the various safety devices, most of which are external to the plant to be protected and are connected by pipework, fitted with taps or cocks for testing purposes. Incidentally my considerate firm, forbade the latter term in reports, out of deference to the typing department!

The simplest safety device is the water gauge fitted to most types of steam boilers. Here the level of water within the boiler repeats within an external glass tube or plate glass assembly and the glass is of heat resistant material. Nevertheless it is a legal requirement that an approved form of protector is fitted, this safeguards personnel should the glass shatter.

The glass can break through incorrect fitting or simply old age, a prudent precaution is replacement at intervals dependent on service conditions, on one of the ships the writer served on the interval was two weeks, glass splinters flying around the confines of the engine room are not known to correct myopia!

The test carried out on these devices is to ensure that all pipework and internal orifices are clear of scale and debris and that the handles of the taps are in the correct orientation to give clear indication that they are either fully shut or fully open.

My first check is to see that the tap handles are in the correct position, top or steam tap fully open, bottom or water tap fully open, drain tap fully shut and no leakage as this can over time drag sediment into the device.

The modern practice of piping all drain discharges out of sight for cosmetic reasons prevents this vital check. The inspection is therefore suspended whilst my esteemed client sorted his problem out - popular job is mine!

The comedy of errors is now about to start as I request the client to test the water gauge, this procedure is strictly hands off for me and every other surveyor, the rule is touch nothing - reason to be given later.

Generally the correctly performed test starts by fully opening the drain tap, steam and water should issue issue forth. The lower or water tap is then shut and if all is well the issue continues, proving that the top piping and orifices are clear. A similar test is carried out by shutting the top tap, and opening the bottom tap, proving the lower piping and orifices are clear.

Finally the top tap is opened and the drain tap closed – simple. As a point of interest if you see this done, say at a heritage site, note the water level before the test and again after the test - it will be higher. Before the test the water in the glass will have cooled over time, become denser and sunk lower, after the test, this water will have been replaced by very hot boiler water, less dense and rising higher. Useless information? Not really, ask me later!

So how come my normally low blood pressure rises exponentially when I am merely an observer? Forget the boiler, it's me that needs the safety valve! For starters since some clients seldom carry out these vital tests, the escaping steam and noise scares the life out of them, with great timidity the drain tap is partially opened, and with a questioning glance they look in my direction, hoping for the affirmative nod that would signal completion of the test - no such luck!

With dramatic gestures I signal that the tap needs to be fully opened, in response, our hero retreats from the gauge to arm's length and gingerly fully opens the tap, now totally enveloped in steam he rushes across and informs me that the test appears satisfactory and asks will it be all right to shut the tap.

With surely commendable diplomacy and bottled up impatience I inform him that the test has not yet even started and fully brief him on the sequence of events required. In response he re-enters the storm cloud, fiddles with the taps which are now barely visible and after much schooling gets it right. As I prepare to leave he asks when my next visit is due, presumably for him to make suitable holiday arrangements!

Boilers and other pressure vessels have devices to control fluid levels and also to give warning of incorrect levels, the latter usually linked so as to cause a total shut down if actuated.

These devices are usually contained in an external float or probe chamber again connected by pipework with taps or valves. The testing of these devices is similar to that for water gauges. The drainage or filling of the chamber should have an observable result, levels within the boiler or vessel should respond to the level within the chamber, alarms should sound for incorrect levels and total shut down should occur if danger is detected.

After testing it is imperative that the various taps or valves are returned to the correct working position, failure to do so can cause a serious mishap and perhaps explosion. If I operate the controls then any subsequent mishap of whatever cause places me in the front firing line, hence the hands off rule. In any case all I am required to do is to witness the tests and report the findings.

The wisdom of this rule was demonstrated at a local factory that the author attended, the boiler attendant tested the devices and supposedly reset them for normal working, whilst I was at home later the same day the manager phoned to say that after my visit the boiler was totally wrecked following a low water condition and required an explanation. Fortunately the attendant confirmed that nothing had been touched by me and that he had failed to reset the controls correctly.

You will have gathered by now that high on my list of priorities is an overwhelming desire to watch my backside, the clients have the numerical advantage when it comes to witnesses, their word against mine!

***

31. Dyeing, defaulting and deception

A new client requested inspections on four dyeing vessels which he had purchased from closed factories with the object of selling them on to foreign buyers. The inspections were required to give potential buyers confidence in the condition of the plant. With the demise of the home textile industry this form of service was much in demand.

This type of plant has been partially described in a previous chapter, a little further description is now required with respect to the support of the vessels. Generally, three supports or legs are attached to the shell. The method of attachment varies and in this case fabricated pocket like fastenings were welded to the shell. The actual legs or supports entered the open pockets from below and were a close fit with the weight of the vessel keeping the legs secure.

The inspection consisted of examination of the shell and ends, including the lid securing arrangements and the various safety devices – all were found to be in good condition.

As is usual in examining second hand plant, an hydraulic test was requested. This was carried out several days later and after being on test for about half an hour, a careful examination was made. All was well except that drops of water appeared around the pocket openings and started running down the legs. This occurred in all four vessels and it was obvious that the shells were cracked in the immediate vicinity of the pockets.

This is a phenomenon we are familiar with and is caused by chemical action. When dye colours have to be changed, thorough cleaning of the vessels is required to remove all traces of previous dye. The chemicals used very often contain caustic or chlorides and during cleaning the liquids occasionally spill over the top of the vessels, run down the side and over the pockets.

Because the legs are a reasonable fit in the pockets, liquid is drawn into the pocket by capillarity. When the vessels are returned to service, heat from the dyeing process evaporates the liquid thereby concentrating it. Over time this cycle of cleaning and usage produces a very high concentration of chemicals.

These chemicals in concentrated form aggressively attack stainless steel when it is in tension as when the vessel is pressurised. The attack is in the form of microscopic cracking likened to crazy paving and eventually as in this case penetration through the full thickness of metal. Undetected this can and indeed has led to explosion. If you have followed this so far – my congratulations!

The repairs consisted in cropping out the defective pieces of shell and pockets thereafter replacing with new plate, electrically butt welded to the parent plate followed by ultrasonic test to check the integrity of the welds.

Our client was not too happy with all this and conveniently went bankrupt before paying us – we had worked for nothing.

About a year later I was called to other premises to carry out similar inspections, the owner turned out to be the same person, trading under a different name, now apparently solvent and prospering!

This time it was money up front or no inspection, the latter prevailed.

***

32. Blow-down and bafflement

EVERY steam boiler has a blow down valve, situated at the lowest part and is used for emptying purposes and also for removing a limited amount of water to correct over high density.

As water is evaporated to form steam, its density rises as chemicals and solids remain. Chemicals are added to maintain the correct alkalinity and mop up dissolved oxygen amongst other things.

Depending on usage or chemical analysis the frequency of removing a limited amount of water varies from daily to extended periods.

The discharged water is at full boiler pressure and therefore considerably above the normal atmospheric boiling point, hence any uncontrolled release to the atmosphere produces an enormous volume of steam.

At large industrial installations the discharge is first piped to large tanks suitably vented and when cooled down is released to the local sewer system. It is illegal to discharge very hot water into sewers.

The writer once attended a small factory producing fire lighters where a boiler and several jacketed pans were under our inspection. I requested the owner to open the boiler blow-down valve to check its working, he asked that I waited a few minutes whilst he checked out the ladies toilets.

This baffled me and I awaited his return with interest. In due course he returned, tested the valve which was fine and explained his absence. With the utmost seriousness he informed me that the discharge piping went to the main sewer but on the way it joined up with the toilet waste pipe.

Whenever he thought his staff were malingering in the toilets he would open the boiler blow-down valve, the arrival of steam in the toilet pans would herald a quick exit back to work!

On the same subject a seagoing friend of mine recalled a visit to a ship repair yard. Here the toilets consisted of a 30' length of 24" diameter steel pipe, pierced along the top with 12" diameter holes, with two bars welded to each hole, door-less partitions were provided. At one end was a large water tank, at set intervals the tank sent flushing water down the pipe. A favourite trick of apprentices was to light a bonfire of newspaper in the first partition, set the flushing operation and wait for the screams of the barbecued!

In the spirit of the day would-be patrons were met at the entrance by an attendant who ensued that time cards were stamped with entry time, on departure cards were stamped again. At the end of the week the time office calculated time spent and made deductions!

Without the slightest doubt these practices will have come to the attention of the Health and Safety people who can be guaranteed to take all the fun out of industrial life!

***

33. Be careful where you sit for lunch

AT a small engineering works one item was under inspection, a vertical air receiver connected to an air compressor. It supplied air to a range of pneumatic tools.

Countrywide we must have had thousands of these simple air vessels under inspection, seldom did they give us any trouble. When atmospheric air is compressed the water vapour present arrives in the receiver, condenses and accumulates at the lower parts, periodically it can be removed through the drain tap.

This condensation corrodes the metal shell which only after many years starts to become a safety issue, occasionally we find a crack around the leg attachments, more about that in another chapter, but otherwise few problems.

It was lunchtime and several of the staff were seated on benches around the receiver and the associated compressor was idly ticking over in an adjacent room. Suddenly there was deafening bang and a sheet of flame erupted from the vessel top, the seated staff ran outside into the road like greased lightning!

Examination found that the large brass safety valve on the receiver top had disappeared through the roof, this explained the sheet of flame. Several months later it was found by a neighbour 300' away in his back garden. It was obvious that an internal explosion had occurred and the only possible cause was ignition of oily vapour. The compressor and pipe to the receiver were examined and the pipe was found to be almost blocked with oily sludge. As for the compressor it must have been well worn at the cylinder bores and piston rings to create the sludge and oily vapour. It had also suffered a broken outlet valve and the breakage had initiated a spark, hence the source of ignition.

There is much to be said for ancient British made

engineering goods. Most of them, compared with modern and foreign manufactured ones, were grossly over engineered and can withstand abuse. Had this been a modern receiver it might well have exploded with fatal consequences.

***

34. Brand new – only fit for scrap

AT a textile mill the client requested that a brand new horizontal air receiver be put on the inspection schedule, it was a self-contained unit with the associated compressor and driving motor mounted on top. This the standard type of equipment found at almost every garage in the land.

At this stage it is normally only a cursory inspection merely to get details for rating purposes and verification of the nameplate details, such as safe working pressure and British or European standard of construction.

Whilst obtaining the plate details I glanced down at the two front legs which attached the unit to the floor, interestingly, the attachment welds for the legs to the shell should have been carefully contoured and blended to the shell to offset the possibility of fatigue cracks forming. This requirement is present in the standard to which this item was supposed to be built under.

However no such essential work had been done but no cracks were visible. Attention was then turned to the rear legs which being against a wall made inspection difficult. With my trusty dental mirror and essential wire coat hanger the attachment welds were finally checked only to find a 2" line of oil smudge present at one attachment the unmistakable sign of the shell being cracked right through.

This is a highly dangerous situation as at any time the crack could have rapidly extended and caused an explosion. The mill engineer was advised to shut the unit down and release the pressure which was promptly done.

The reciprocating compressor mounted on top of this type of unit can, if not well balanced, furnish the vibration necessary to start a crack in an imperfect weld.

Defects of this nature in this type of plant are not worth repairing and the only solution is replacement.

Obviously I cannot disclose the make of this unit but I can say it was manufactured at the Western end of Europe and its citizens spend most afternoons asleep!

***

35. Jacketed Pans and exotic hams

A PAN in a dinner jacket is definitely not the subject of this chapter but refers to a device for heating liquid products by means of steam. The pan containing product is enclosed in a similar outer vessel, steam occupies the space between the two and its heat transfers to the product. The temperature attained can be precisely controlled by adjustment of the steam pressure, a given pressure equates to a definite temperature as per the laws of physics.

Pans of various types and sizes are used throughout industry but mainly in food preparation and chemical processing. The most common form is a hemispherical product pan of relatively thin copper, fitted into a cast iron jacket, the copper pan is flanged at the top and surmounted by a clamping ring bolted or studded to the cast iron. The device is on gimbals or trunnions to enable the pan to be decanted and the product removed.

A more modern version of approximately the same shape is made entirely of stainless steel, all welded and hence cannot be dismantled. In both types steam enters through a hole in one gimbal and condensate exits the same way through the other gimbal.

Other designs are of vertical cylindrical construction jacketed on the shell side only and not on the bottom plating. I must mention one other type the 'Bain Marie' as this enables me to demonstrate a degree of scholarship, the translation is 'Bath of Mary' It is to be found in large kitchens and is in fact a rectangular bath about 3" deep containing boiling water, cooked food in pans or dishes are placed within and are kept hot till required.

How the lady in question survived in a bath of boiling water is beyond me, hopefully some reader may be able to throw some light on this hot subject.

This type of plant together with a myriad of others is defined in law as a steam receiver and must have periodic thorough examination. The inspection focus is on the external surfaces of the inner pan which through wear and tear deteriorates, the steam supply arrangements with particular attention to the capacity and condition of the associated safety valves. With copper pans an occasional inspection when dismantled is requested as fatigue cracks sometimes appear at the underside of the top flanging. Generally we request a hydraulic test every four years or so on most types of pan due to the impossibility of steam side inspection.

The writer once attended what can only described as a back street food processor producing an exotic range of cooked meats. Entry to the works revealed a scene reminiscent of a TV period drama – a medieval castle kitchen, a row of steaming cauldrons attended by a kitchen hand replete in leather apron and clogs. Like an automaton he moved from pan to pan, momentarily stopping for a quick stir with something like a rowing boat oar.

My inspection was limited to one jacketed pan which was cold and empty, whilst leaning over, something wet and warm hit me on the back of the neck, looking upward I saw a large skylight dripping with condensation caused by the rising steam from the other working pans.

The glass was filthy, yellowed with fly droppings, festooned with cobwebs and the odd bluebottle for good measure. I carried on my inspection but decided that I would give cooked meat a miss for the time being, my decision hardened when a bluebottle in its funeral shroud of spider web took its final plunge into the gravy grave of the adjacent working pan.

The embalmed bluebottle floated high on the gravy sea as per Archimedes laws of floatation, giving me time to summons the aproned kitchen hand. He merely shrugged shoulders and with a deft flick of the wrist used his oar to inter the corpse.

Next time at the supermarket, be adventurous, look out for calliphora ham (bluebottle), you now know its provenance!

***

36. The Thin Yellow Line

FOUNDRIES were often on my inspection schedule as they are users of both carbon dioxide and compressed air, both gases require storage at pressure and hence inspection of the containment vessels.

The inspection of one such carbon dioxide vessel was scheduled for a Saturday morning when the foundry would be shut. Having met the owner I followed him through the quiet and gloomy works as he led the way to the vessel. Out of habit and a throwback from sea going, as my torch was illuminating the various machines being passed - none of which were under our inspection - it picked up an elderly two cylinder reciprocating air compressor motor driven via belts and flywheel, stationary now as it was the weekend.

Interesting! In the odd second or two as I passed and in marine engineer style my torch started at the cylinder head, the unloader device, crankcase doors, foundation bolts and flywheel – an inspection on the hoof! At the protrusion of the crankshaft from the flywheel a distinct yellow brown stain was visible – indisputable evidence that the flywheel was slack on the shaft, the harbinger of breakdown and expense. The owner was advised and no doubt to his surprise he found the securing bolts and key quite slack as he wielded spanners and hammer.

The reliability of mechanical plant depends on many aspects including the requirement that where adjacent parts are rigidly held together, they are maintained in that state. Any relative movement causes wear and accelerating slackness and ultimately failure. With ferrous metals such as steel or cast iron such as in this case, the wear produces microscopic shards of metal which over time migrate to the exposed junction edges, the clean shards quickly rust on exposure to the humidity of the atmosphere, hence the thin yellow line.

The carbon dioxide vessel, the real reason for the visit, was in good order, I have never found a problem with these vessels, many are spherical and like the soap bubble are perfect for the containment of fluids under pressure.

***

37. Trainees, traumas and tantrums

LIVING near my head office it was convenient for trainees to be sent to be with me for up to three weeks at a time for field training and over the years I must have had over thirty to accompany me.

This of course greatly slows down the rate of work but is hugely compensated by the fact that by becoming a teacher, then by your pupils you'll be taught. In a job that is performed on one's own it is easy to slip into inappropriate practices and inefficient working, questioned by trainees the areas for improvement are soon revealed.

I had arranged that on the last Friday of the town wakes fortnight to attend at the factory of a new client to inspect a large steam boiler, with a trainee in tow we called at the chief engineer's office who arranged for us to be taken to the boiler house.

On arrival it was obvious that no attempt had been made to prepare the boiler for inspection, none of the doors and manholes had been opened up, no fittings were apart. I assumed that probably some mistake had been made with the date for the cleaning contractors to attend, so we made our way back to the engineer's office to explain the situation and seek another date for inspection.

On hearing that we were unable to even contemplate an inspection the engineer went totally berserk, yelling and shouting that if he arranged for my attendance then come what may I would carry out the required inspection, in his dreams! Each sentence was punctuated by a tremendous fist thump upon his desk – you certainly meet some strange people in this job!

The desk had a large Victorian glass inkwell upon it, my interest switched to this object as it slowly traversed the desk top, no doubt in compliance to some aspect of Newtonian mechanics which I found unfathomable. Its course suggested that with a few more thumps it would reach the desk edge, from thence to Newton's and my delight it would drop over the edge and discharge its ink cargo over the pristine carpet.

No such luck, it slightly overlapped the edge but its centre of gravity firmly remained over the desk and it stubbornly refused point blank to go any further. Our man was still shouting and thumping. We, like disappointed, theatre patrons decided to leave the auditorium and quietly left the premises. This was a real eye opener to my colleague, but I assured him that this was a new experience for me as well.

It was only nine thirty in the morning and I had the rest of the day to occupy the trainee, it was a beautiful summer's day with a light wind – Sailing weather!

About a year previously I had built a 14' sailing dinghy and joined a local club, nobody at my head office had specified the exact nature of the training, only that my colleague was to accompany me and get on the job training. To my mind after a set too with an esteemed client the best course was a cooling down period, hence we made our way up to the club.

Three hours later he had acquired two new skills, how to deal with a difficult client and how to sail a small boat – a highly profitable day! A quick trip to the rail station and he was off to his home in Wales. The following week could be interesting....

After picking him up at the station we spent Monday and Tuesday carrying out the 'Working Inspections' of plant examined at several premises the previous week, described in another chapter.

On Tuesday night whilst watching the television the doorbell rang, the visitor introduced himself as a director of the factory which we had visited on the Friday. He profusely apologised for the behaviour of the engineer whom he had now sacked. He explained that the factory was shut pending the examination of the boiler and that he had arranged for a contractor who had completed the necessary preparation, therefore could I attend later in the week.

I could do better than that, I offered to return with him that same evening and carry out the inspection, by midnight the job was done, enabling the factory to start the next day.

I am fortunate in having a social worker to hand, my wife, to whom I can discharge the pressure developed by the odd client! I am given to understand that in the case of this engineer insecurity was his problem, linked to some deficiency in his early toilet training. This analysis, like most other social work explanations, was completely beyond me!

***

38. Two Sea Lawyers meet up for a chat

MOST inspections of minor plant are initiated by a phone conversation, the client is reminded that an item is due for inspection and a mutually convenient date and time is arranged.

It was unusual that a client wrote to my Manchester office stating that his air receiver could be inspected at any time. The client was not in my district but as the usual surveyor was on holiday the letter came to me and the formality of it was surprising.

I duly called at the clients' premises which happened to be a garage, met the owner and was conducted to the receiver. This was a small horizontal type with a compressor mounted on top, standard garage equipment for amongst other things inflating tyres.

Neatly arranged on the floor was a paint tin full of paraffin, a pile of rags, several spanners and two spare joints for the receiver inspection doors.

I informed the esteemed client that when he had opened up and cleaned the receiver I could then carry out the inspection. On the wall behind was a notice board containing Factory Act particulars, drawing attention to the occupiers' responsibilities – this information board is to be found in all industrial premises.

With a greasy finger he pointed to the paragraph relating to air receivers and attempted to explain that it was I who should prepare the item for examination. I began to think that that perhaps all this was some hoax and maybe next week I would appear on 'Candid Camera'!

He was quite serious however and despite my diplomatic interpretation of the relevant Act I failed to convince him that he as the occupier was responsible for the preparation after which I would carry out the inspection. In close on 30 years in the inspection business, he was the first person to hold this view, the fact that he had prepared for the visit with an array of cleaning materials and spanners suggested he was also carrying a large chip. Even my social worker wife was baffled, the usual hot chestnut of interrupted toilet training or too long in nappies did not appear to explain his behaviour.

Whilst mulling the situation over, the solution was staring me in the face – the notice board,

How about getting the local factory inspector to come down and interpret the law for us? Much to my surprise and relief he swallowed the bait whole, totally ignorant that I knew something completely unknown to him.

The inspector was known to me as a proverbial tigress, two piece suit crowned by a feathered hat. She would certainly sort out his air receiver problem as the starting point, thereafter the rest of the day would be spent in a minute inspection of the whole of his premises, starting with the vehicle inspection pits and ending up somewhere near the roof.

After seeking his confirmation that this was the path he wished to take I promised him that once home the phone call would be made, I left the garage feeling a little guilty, two weeks later the client phoned me, a changed man, now contrite, courteous and extremely cooperative.

He explained that the garage had been shut down for two weeks as he endeavoured to comply with the huge list of matters at variance with the Act. The largest job was the provision of proper washing and toilet facilities for his staff, followed by a rewire, proper lights in the pits, whitewashing the whole garage, attention to a myriad of minor infringements and, by the way, the air receiver had to be cleaned for inspection.

I called round and was amazed at the transformation, the receiver appeared to have been steam cleaned and valeted, the brass nameplate together with the safety valve had had a good lick of Brasso, a box of clean rags appeared on the floor presumably to wipe my hands on and an invitation was given to visit the new wash house.

It was my guess that this client would give no more trouble!

***

39. Finished with Engines

THE voyage is over, the destination reached, the engine room telegraph rings out and indicates ' Finished with Engines'.

For the engine room staff the work changes, main boilers to be shut down, forced and induced draught fans to shut down, donkey boiler to be fired up, steam to the winches, engine turning gear to be engaged, steering gear to be shut down, etc.

Lloyds' engineer-surveyor due tomorrow noon, survey of No.2 Diesel Generator crankshaft. Afterwards it needs to be boxed up and running so that the ship can sail on the next high tide, any delay and the sailing will be postponed till the next tide twelve hours hence, ship owners' don't want excuses but profit!

As a first tripper so far as authorship is concerned, the voyage is over, finished with chapters has arrived, it's closing down time, checking grammar and spelling, hand over to a publisher, return to my cabin and think about the next book...

THE END

****

Printed in Great Britain
by Amazon

83052632R00078